T0235430

Lecture Notes in Computer Science   **12567**

More information about this subseries at http://www.springer.com/series/7409

Xiaofeng Meng · Xing Xie ·
Yang Yue · Zhiming Ding (Eds.)

Spatial Data and Intelligence

First International Conference, SpatialDI 2020
Virtual Event, May 8–9, 2020
Proceedings

 Springer

Editors
Xiaofeng Meng
Renmin University of China
Beijing, China

Xing Xie
Microsoft Research Asia
Beijing, China

Yang Yue
Shenzhen University
Shenzhen, China

Zhiming Ding
Chinese Academy of Sciences
Beijing, China

ISSN 0302-9743   ISSN 1611-3349   (electronic)
Lecture Notes in Computer Science
ISBN 978-3-030-69872-0   ISBN 978-3-030-69873-7   (eBook)
https://doi.org/10.1007/978-3-030-69873-7

LNCS Sublibrary: SL3 – Information Systems and Applications, incl. Internet/Web, and HCI

This Springer imprint is published by the registered company Springer Nature Switzerland AG
The registered company address is: Gewerbestrasse 11, 6330 Cham, Switzerland

Preface

This volume contains the papers from the ACM SIGSPATIAL China annual conference (Spatial-DI 2020). The event was organized by Shenzhen University, and the conference was themed 'Spatial Intelligence: Convergence, Convergence and Productivity'.

ACM SIGSPATIAL China is committed to promoting the research paradigm of spatial data and the innovation and application of spatial intelligence theory and technology in the fields of space-time big data, smart city, traffic science and social governance.

Due to the impact of COVID-19, the meeting scheduled for March 27–28 in Shenzhen was replaced by a webinar. The meeting was held from May 8th to 9th 2020, while a special forum for invited guests was held on May 8th, and a special session for receiving the report of the paper on 9th May.

Despite the online format the meeting still maintained a high level, with well-known scholars and Internet industrial companies such as Baidu, Alibaba, DiDi, Huawei and JD represented in the intelligent acquisition, spatial data management, analysis and application aspects of the workshops.

Spatial-DI 2020 mainly aimed to address the opportunities and challenges brought about by the convergence of Computer Science, GIScience, AI and more. The main topics of the conference included: Spatial Machine Learning and Artificial Intelligence, Spatial Data Acquisition and Positioning, High-Performance Computing of Large-Scale Spatial Data, Mobile Data Management and Analysis, Geographic Information Retrieval, Spatial Semantic Analytics, Autonomous Transportation and High-Precision Maps, Urban Analytics and Mobility, Spatial-Temporal Visualization and Visual Analytics, Location-Based Services and Privacy Issues, Geo-Social Network Analytics, and Geo-computation for Social Science.

The proceedings editors wish to thank the dedicated Scientific Committee members and all the other reviewers for their contributions. We also thank Springer for their trust and for publishing the proceedings of Spatial-DI 2020.

Xiaofeng Meng
Xing Xie
Yang Yue
Zhiming Ding

Organization

Scientific Committee

Xiaofeng Meng	Renmin University of China, China
Feng Lu	National Remote Sensing Center of China, China
Xing Xie	Microsoft Research Asia, China
Yang Yue	Shenzhen University, China
Zhiming Ding	Institute of Software, CAS, China
Guangzhong Sun	University of Science and Technology of China, China
Huayi Wu	Wuhan University, China
Danhuai Guo	Computer Network Information Center, CAS, China
Beihong Jin	Institute of Software, CAS, China
Hui Zhang	Tsinghua University, China
Yan Huang	University of North Texas, USA
Li Xiong	Emory University, USA
Fusheng Wang	Stony Brook University, USA
Ralf Hartmut Güting	University of Hagen, Germany
Christian Jensen	Aalborg University, Denmark
Xiaofang Zhou	The University of Queensland, Australia
Yanbo Han	North China University of Technology, China
Bin Cui	Peking University, China
Guoliang Li	Tsinghua University, China
Guangyan Huang	Deakin University, Australia
Amr Magdy	University of California, Riverside, USA
Bruno Martins	Instituto Superior Técnico, University of Lisbon, Portugal
Mark McKenney	Southern Illinois University Edwardsville, USA
Anirban Mondal	Ashoka University, India
Anand Padmanabhan	University of Illinois at Urbana-Champaign, USA
Dimitris Papadias	The Hong Kong University of Science and Technology, Hong Kong SAR, China
Kostas Patroumpas	Information Systems Management Institute, Athena Research Center, Greece
Nikos Pelekis	University of Piraeus, Greece
Dieter Pfoser	George Mason University, USA
Jeff Phillips	University of Utah, USA
Gianluca Quercini	CentraleSupélec – LRI, France
Chiara Renso	ISTI-CNR, Italy
Dimitris Sacharidis	Vienna University of Technology, Austria
Mohamed Sarwat	Arizona State University, USA
Markus Schneider	University of Florida, USA

Monika Sester	Institute of Cartography and Geoinformatics, Leibniz Universität Hannover, Germany
Bettina Speckmann	Eindhoven University of Technology, The Netherlands
Rade Stanojevic	Qatar Computing Research Institute, HBKU, Qatar
Sabine Storandt	University of Konstanz, Germany
Gautam S. Thakur	Oak Ridge National Laboratory, USA
Yannis Theodoridis	University of Piraeus, Greece
Kristian Torp	Aalborg University, Denmark
Martin Werner	DLR Oberpfaffenhofen and Leibniz University Hannover, Germany
Stephan Winter	The University of Melbourne, Australia
Raymond Chi-Wing Wong	The Hong Kong University of Science and Technology, Hong Kong SAR, China
Jianqiu Xu	Nanjing University of Aeronautics and Astronautics, China
Bin Yang	Aalborg University, Denmark
Demetrios Zeinalipour-Yazti	University of Cyprus, Cyprus
Baihua Zheng	Singapore Management University, Singapore
Andreas Züfle	George Mason University, USA
Wei-Shinn Ku	Auburn University, USA

Organizing Committee

Wei Tu	Shenzhen University, China
Zhensheng Wang	Shenzhen University, China
Jizhe Xia	Shenzhen University, China
Jincai Huang	Shenzhen University, China
Jinzhou Cao	Shenzhen University, China
Jin Zeng	Shenzhen University, China
Tianhong Zhao	Shenzhen University, China

Contents

Visualization Science

Traffic Management

A Fast Clustering Approach for Identifying Traffic Congestions

Yan Zhuang[1]($\boxtimes$), Cheng-Lin Ma[2,3], Jin-Yun Xie[3], Zhui-Ri Li[3], and Yang Yue[3,4]

[1] State Key Laboratory of Information Engineering, Mapping and Remote Sensing,
Wuhan University, 129 Luoyu Road, Wuhan, Hubei, China
zhuangyan@whu.edu.cn
[2] Guangdong Key Laboratory of Urban Informatics, Shenzhen University, 3688 Nanhai Avenue,
Shenzhen, Guangdong, China
[3] Shenzhen Key Laboratory of Spatial Smart Sensing and Services, Shenzhen University,
3688 Nanhai Avenue, Shenzhen, Guangdong, China
[4] Department of Urban Informatics, Shenzhen University, 3688 Nanhai Avenue, Shenzhen,
Guangdong, China

Abstract. DBSCAN, a density-based clustering algorithm, has been widely used in pattern recognition and data mining. However, under large-scale streaming data scenarios, it suffers heavy computational cost because it examines distances between each points multiple times, especially in traffic applications which usually require calculating road network shortest distance instead of Euclidean distance. Therefore, the performance of DBSCAN for real-time clustering analyses is has become a bottleneck in such applications. Focusing on fast identifying traffic-related events, this paper utilizes linear feature to improve the efficiency of clustering by introducing linear referencing system (LRS). LRS has long been used in managing linear features, which could simplify shortest-path computation into 1-dimensional relative distance calculation, thus can significantly reduce computational complexity and cost, and meet the real-time analysis requirement of streaming data. Using vehicle GPS trajectory as an example, this study designs a LRS and its associated dynamic segmentation method for identifying traffic congestions. Experiment results proved the flexibility and efficiency of the proposed LRS-based clustering approach in identifying traffic congestions.

Keywords: Clustering · DBSCAN · Linear trajectory · Linear referencing system · Traffic congestion

1 Introduction

DBSCAN (Density-based spatial clustering of applications with noise) [1] is one of the most widely used density-based clustering algorithms. Based on the notion of density reachability and the minimum number of points (*minPts*), DBSCAN defines clusters by iteratively examining the distance ε (*eps*) of a point p with its neighborhood points. Therefore, calculating distance is a major cost, especially in transportations related applications in which road network-based shortest distance is often required rather than Euclidean

X. Meng et al. (Eds.): SpatialDI 2020, LNCS 12567, pp. 3–13, 2021.
https://doi.org/10.1007/978-3-030-69873-7_1

distance, because the computational complexity of shortest-path algorithms for obtaining network distance, such as Dijkstra or A*, is much higher than that of Euclidean distance.

For example, Vehicle GPS trajectory data is one of the most studied spatiotemporal data, and DBSCAN is a straightforward approach to identify traffic congestion or other abnormal conditions using this type of data source. However, to identify congestions on the fly using real-time GPS data stream, DBSCAN has encountered the efficiency barrier, and a more efficient clustering algorithm is necessary in addition to a large-scale computing framework. Under this circumstance, this paper introduces linear referencing system (or linear reference system, LRS) to improve the clustering efficiency by reducing distance computation complexity. LRS is an alternative to manage locations of linear features, which express geographic locations (x, y) by relative positions along a measured linear feature, for instance, *route* G4-100 + 23.2 km (route ID –mileposts + offset), rather than at a GPS coordinate. With LRS, locations of vehicles or events along road links can be referred to by their distances from a known point; and thus, the calculation of shortest-paths along road networks can be simplified into obtaining 1-dimensional relative distances. The key issue is how to design the LRS and transform the linearly referenced data into map locations for further use, i.e., dynamic segmentation.

To improve the efficiency of DBSCAN, maintaining a distance matrix is a commonly used strategy to avoid repeated distance calculation even for Euclidean distance. However, this strategy is not applicable for streaming data scenario in which the distances keep changing with the generation of new data points; and thus, it is meaningless to maintain such a distance matrix in real-time manner.

The contributions of our work are summarized as follows:

- We proposed a fast clustering algorithm for identifying linear events by incorporating LRS.
- We designed a LRS and associated dynamic segmentation method for vehicle GPS data stream, which can be applied in other related applications.
- The LRS-based clustering approach could improve the accuracy of identifying traffic congestions by avoiding some false results generated by DBSCAN.

The rest of the paper first designs a LRS and dynamic segmentation method for real-time and large scale vehicle GPS data management, and then discusses the LRS-based clustering approach. Experiments and results are given to illustrate the effectiveness of the proposed approach.

2   Related Studies

To deal with the continuous feature of streaming data, various incremental strategies have been proposed to avoid starting from scratch in every clustering process. The general idea is managing data in a predefined structure based on data summarization or pattern mining [3–6], which usually works with an online data processing and an offline data maintenance parts [7, 8]. For example, [4] utilized representative points as a knowledge repository to assist new data processing and data archival. Other structures, such as cuboid [5], PrefixTree+ [6], tree [7], cell tree [9], and graph [10] were also used to maintain clusters and process evolving data. Most of these structures were designed for genetic type of data, and can be applied on both free space 2-dimenional (x, y) and 3-dimensional (x, y, z) trajectory data, for instance, animal or hurricane trajectories. Therefore, the computation for distance is one of the major cost.

Different from free space trajectories, vehicle and most human trajectories in urban areas are constrained by road networks, and calculating the distances for vehicle GPS trajectory data can be simplified into 1-dimensional linear space. Thus, the computational efficiency could be improved as long as because LRS can turn the computation of shortest-paths into calculating 1-dimensional relative distances.

Studies on linear trajectory clustering [3, 8], however, have not fully used of this characteristic.

LRS is an intuitive way to represent linear distance and an effective approach to simplify the distance computation. The next section will first introduce LRS designed for vehicle GPS trajectory data management and clustering.

3 LRS for Managing Vehicle Trajectory Data

First, some terms are defined to describe the LRS, referring the conventional GIS definitions [11]:

Definition 1: "**Link**": the basic component of a route which has a unique identifier. A link has a start node, an end node, and with or without shape points between the two nodes to define its geometry.

Definition 2: "**Route**": a route is a collection of links, which has a unique identifier.

Definition 3: "**Event**": an event is a point or linear feature that occurs at or along a link or a route. Examples of a point event in transportation fields might be accident sites, and linear events might be speed limits and congestions. Events are stored in event tables.

Definition 4: "**Dynamic segmentation**" is a process of computing the map locations of events stored and managed in an event table using a linear referencing system.

It is "dynamic" because each time an event changes along routes, the associated links need not to be spitted physically. Using dynamic segmentation, various linear features can be displayed, queried, edited, and analyzed without affecting the underlying linear features' geometry.

3.1 Creating LRS

Adopting LRS is a common practice in transportation networks, rivers, pipelines, and sewer networks management. This is a process of encoding a location. A well-known LRS-based application is OpenLR [12] which provides a dynamic location referencing method for data exchange and cross-referencing using digital maps of different vendors and versions. The core of OpenLR is using pre-coded locations which are then added to the corresponding maps. In a board sense, the pre-coded locations used in OpenLR is a

LRS. Besides road name or intersects, milestone system is the most widely used linear referencing method, such as *route* G4-100 + 23.2 km.

This study created a layer of linear-referenced-points (LRPs) as virtual milestones. The numbered round dots in Fig. 1 are the LRPs which segment a link with regular distance, while the triangle shapes on the links are the vehicle GPS points (after map-matched) associated with instant speed *Vmi* and time stamp *Tmi*.

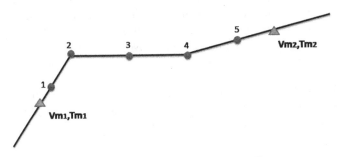

Fig. 1. Linear-referenced-points (LRP)

In this study, the distance between the LRPs is determined based on the average length of road links. It could also be adjusted according to applications. In our dataset, the distance was set to 100 m since the average link length is about 270 m, and most of the vehicle GPS data were sampled around every 60 s.

Creating the layer of LRP increases the size of road network files. The increased size depends on the selected distance interval, also the straightness of the road links. Except straight road links which only have two shape points (a start node and an end node), others road links have multiple points stored in road network database to define the geometry of the links. For example, the 2# and 4# dots in Fig. 1 are the shape points. In some road networks with relative straight routes, the shape points could be used for linear referencing; however, the shape points sometimes are redundant, sometimes are sparse, which makes LRS inefficient. This is the reason that why we created the virtual LRP layer to overcome this problem. In our study dataset, the number of road links of an urban is 21,114, and number of shape points is 142,932, while the number of virtual LRP is 113,497. The LRP file can be easily stored in memory.

3.2 Dynamic Segmentation

The original aim of dynamic segmentation is locating events onto the map based on LRS, this is a decoding process. However, since map-matching is often necessary for vehicle GPS data, the matched GPS coordinates are "distorted" coordinates of the vehicles and only represent the approximate locations of vehicles, instead of the accurate vehicle positions. Therefore, we discarded the matched locations, and only stores the LRPs instead. The attributes of raw GPS data were transferred onto these LRPs, such as speed and time. The underlying assumption is, the attributes between two consecutive vehicle GPS points are homogeneous.

This average speed between two consecutive vehicle GPS points is assigned to LRPs between them. Figure 2 illustrates this approach, in which (v_i, t_i) represents the corresponding travel speed and its associated time stamp for GPS points and LSPs. Thus, the dynamic segmentation problem turns into calculating the associated travel speed for each LRP.

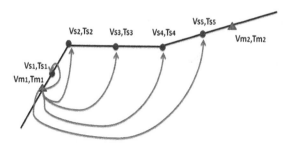

Fig. 2. Calculating LRP travel speeds using two consecutive GPS points.

To assign vehicle speeds onto LRPs, one additional record was added to trajectory data records, which is the distance ratio of the map-matched vehicle GPS points with the length of its matched road link. This could be done during map-matching. Then the table of floating car data includes the following items: {*car_ID, t, (longitude, latitude), speed, pre_Route, current_LinkID, f_Ratio*}, which describes the fact that: at time "*t*", a vehicle ("*car_ID*") passed a road link ("*current_LinkID*") at location "*(longitude, latitude)*" with instant speed ("*speed*"), where the ratio of the traversed distance at "*current_LinkID*" to the total length of the road link is "*f_Ratio*"; and the vehicle has passed a series of road links (*pre_Route*) between this location and its previous point.

With the help of the distance ratio, vehicle travel speeds can be allocated to LRPs with a faster computation of distance:

$$L = \sum_{i=1}^{n} l_i * p_i \tag{1}$$

where, n is the number of road links that the two consecutive vehicle GPS points traversed; l_i is the length of the ith road link; and p_i is the distance ratio of the vehicle GPS point on its mapped road link; $p_i = 1$, for the whole road links between the two GPS points, if any. The assumption is that the vehicle travel with uniform speed between two consecutive GPS records.

Instead of using the distance equation: $L = \sqrt{(x_1 - x_2)^2 + (y_1 - y_2)^2}$, introducing distance ratio can reduce the computation complexity and cost, and thus meeting real-time processing requirement for streaming data.

Then, the travel speed v associated with a LRP can be obtained by:

$$t_i = \frac{\Delta l}{L} * \Delta t = \frac{\sum_{i=1}^{n} l_i * p_i}{L} \tag{2}$$

where Δt is the time difference between two consecutive GPS points.

Each LRP also has a distance ratio item "LR-ratio" which is similar to the "*f_ratio*" of GPS points as explained above. This ratio is mainly used for back-calculating the coordinates of LRPs so that to draw them on the map, since the coordinates of start node and end node of the link are known information in road network data. This table can be further be simplified by generalizing speed information into patterns.

So, the event table stores the created LRPs and their associated travel speeds. Following table shows the structure of the event table:

Table 1. LRP event table

Item	Description
NodeID	LRP ID
LinkID	Associated road link ID
SPEED	{car_ID, t, speed}
LR-ratio	distance (startPt, lrPt)/distance (startPt, endPt)
NodeID	LRP ID

Since the event table can be modified incrementally, it is suitable for streaming data processing and analysis.

4 LRS-Based Clustering Approach

With the event table designed illustrated in Table 1, each LRP has a linked list which store vehicles that passed the point and its associated travel speed and time stamp (ID, v_i, t_i). To identify traffic congestions, DBSCAN visits each slow point (for example, $v < 15$ km) and retrieves its neighborhood points. As above discussed, this procedure demands heavy computational resources for online analysis because it requires multiple times of data retrieval, query and shortest-path calculation. Meanwhile, since the LRP layer has been added and stored with road network data, and the event table were also constructed during map-matching process, identifying traffic congestion can be simplified into a table query procedure based on the event table. Given some thresholds, such as the threshold for slow traffic, examined time period, and the minimum number of slow points that forms the congestion as the *minPts* used in DBSCAN, identifying clusters is just a query.

The essence of the proposed LRS-clustering approach is measuring the level of congestion by counting the number of slow moving vehicles. The underlying assumption is intuitive: the more slow vehicles, the likely a congestion happens. One of the advantages of utilizing LRS is, the congestion road sections could be identified and represented in a finer spatial unit, for example, 100 m, which is more accurate and more flexible than using road link as the basic spatial unit. And adjacent congested small unit could be combined into a longer congestion section.

The overall run time complexity of DBSCAN is $O(n \log n)$ if an indexing structure is used, or $O(n^2)$ without an index. Often the distance matrix of size $(n^2 - n)/2$ is materialized to avoid calculating distance repeatedly. This, however, also needs $O(n^2)$ memory, whereas a non-matrix based implementation only needs $O(n)$ memory. The time complexity of the proposed LRS-clustering is $O(n \log m)$ and it needs $O(m)$ memory, where m is the total number of road links.

5 Experimental Results

5.1 Data

The experiment data were GPS records collected by over 11,000 taxis from Shenzhen, a major city in the south China. Each taxi GPS record contains information of car ID, GPS location, timestamp, spot speed, and azimuth. The collected time interval was around 60 s. The road network data has 21,114 road links covers about 2,000 km^2.

5.2 Computational Costs

In this experiment, we used speed $v \leq 15$ km/h (or 4.167 m/s) as the threshold to define congestions. Choosing the *minPt* in the LRS-clustering is not deterministic. Figure 3 shows the numbers of slow moving vehicles on the LRPs, during a 10 min time interval, which follows the power law distribution. Based on this distribution, we chose the minPts = 10.

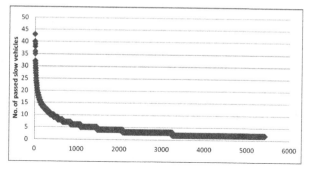

Fig. 3. The number of slow vehicles associated with LRPs (Monday 08:20–08:30).

Under this circumstance, the LRS-clustering approach found 457 congestion points as shown in Fig. 4, when *minPts* = 10 which means a section of a road link is marked as "congestion" if at least 10 slow vehicles passed the LRPs on this section during the 10 min interval.

To illustrate the performance of the proposed LRS-based clustering approach, the computation costs were compared with those of DBSCAN. As it is well known that the determination of the two parameters of DBSCAN is relatively subjective, and with different parameters, the outcomes and computational costs also various. In general, the two clustering approaches are based on different architectures, they are not comparable in terms of computational time, though theoretically the proposed linear clustering approach is much faster than DBSCAN ($O(n \log n)$ vs. $O(n \log m)$ as analyzed above). To

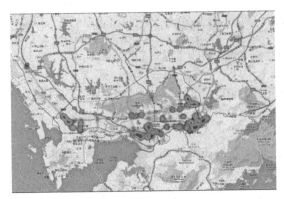

Fig. 4. Clusters identified by LRS-clustering approach (minPts = 10).

compare the performances, we calculated a series of clusters identified by each of the two clustering approaches, and set a benchmark as *minPt* = 100. Since in our experimental dataset, the LRPs were added every 100 m, the *eps* for DBSCAN was also set to 100 m. Most important, we also set the DBSCAN distance threshold to 100 m which means the shortest-path calculation only consider those points within 100 m. The aim is twofold: 1) to reduce the distance computation cost, and 2) to make it more comparable with LRS-based approached. Table 2 shows the performances of the two clustering algorithms, which lists the execution times for both the data initialization and clustering process.

Table 2. Comparison of computational time for LRS-clustering and DBSCAN

Dataset size	Execution time (sec)			
	LRS-clustering		DBSCAN	
	Initialization	Clustering	Initialization	Clustering
42,098	6.458	0.965	6.677	0.764
42,465	6.712	0.837	6.737	0.799
43,003	2.194	0.944	1.979	0.844
84,197	4.722	1.938	4.0764	3.000
84,931	5.465	1.618	4.382	2.910
86,007	4.257	1.722	3.250	2.979
168,397	8.111	3.847	5.181	12.153
169,865	7.903	3.556	4.958	11.930
172,017	8.347	3.486	5.236	11.736
336,795	16.640	6.888	8.945	46.528

(continued)

Table 2. (*continued*)

Dataset size	Execution time (sec)			
	LRS-clustering		DBSCAN	
	Initialization	Clustering	Initialization	Clustering
339,732	16.528	7.333	9.417	49.611
344,035	16.778	7.417	10.278	49.583
673,592	32.388	14.056	12.889	188.389
679,465	33.556	15.056	13.500	214.833
688,072	33.667	14.778	14.389	202.667
1,347,190	64.556	27.555	24.888	846.000
1,358,930	64.333	27.889	26.222	962.000
1,376,150	70.667	28.889	29.111	923.333

It can be observed that, with the increase of dataset, the LRS-clustering approach outperformances DBSCAN dramatically, though DBSCAN has advantage at smaller data size scale. We continue increased the dataset used for clustering, and the execution time of LRS-based clustering can be only about 1/10 of that of DBSCAN.

5.3 Clustering Accuracy

In terms of accuracy, it also should be noted that, comparing with the proposed LRS-clustering approach, DBSCAN has a drawback—it may generate bias due to uneven spatial distribution of vehicles with different sampling time intervals, because the clustering result of DBSCAN is determined by the number of neighborhood points. In our case, the GPS data were collected from several taxi companies in the study area, and each company often has its own data collection specification. In this dataset, some GPS trajectories were collected at 10 s interval while some were collected at 60 s interval, or even longer. Therefore, the regions covered by 10 s GPS time sampling intervals are tend to be detected as "congested" because the GPS point density of these regions may be higher than those 60 s-regions even with similar traffic conditions. For the proposed LRS-based clustering algorithm, GPS points were sampled onto LRPs, so that each vehicle only has one record on each LRP no matter the data collection time interval. This can avoid the bias of DBSCAN. For example, with eps = 100 m and minPts = 200, DBSCAN detected 22 congestion clusters; while LRS found 34 clusters with minPt = 7. Our experiment found that the number of identified congestion clusters by LRS-clustering approach is often larger than that of DBSCAN under the similar condition. This is because what the LRS-based clustering approach identified is the number of congested LRPs, while DBSCAN combines adjacent small clusters into larger ones. But the identified congestion areas and their sizes in general, are similar. Meanwhile, the number of associated speed points on LRPs can represent the level of congestion, while DBSCAN can only represent whether it is congested or not, determined by the given *minPts*. Thus the LRS-based approach can achieve better visualization effect to represent the level of congestion.

错误!未找到引用源。highlights some the differences of the 22 clusters vs. 34 clusters. The red points are those detected by DBSCAN, and those represented by heat map are the results of LRS-approach. Some clusters were overlapped, some were close, but DBSCAN found a cluster while LRS-approach has not (the upper right one). We examined the database, and found that the cluster points were generated by one taxi which recorded GPS data every 10 s. Although without the ground truth congestion data, we could not validate which algorithm is more accurate, it was very likely that the taxi was parking, or waiting for passengers, rather than a real congestion (Fig. 5).

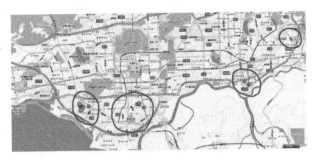

Fig. 5. Differences between identified congestions by DBSCAN and linear clustering approach.

Therefore, using the LRS-based clustering approach is not only a faster clustering algorithm, but also can overcome this disadvantage of DBSCAN.

6   Conclusions

This paper proposed an LRS-based clustering approach which could improve the efficiency of DBSCAN when calculating shortest-paths for network-based trajectory data streams. Since DBSCAN identifies clusters by examining the distances between each points multiple times, and when considering network distance, its computational complexity and cost is even higher. Thus, it is inefficient in real-time manner for streaming data.

The proposed LRS-based clustering approach introduced LRS to avoid shortest-path computation, and thus can reduce both the computational complexity and cost for linear trajectories. Vehicle GPS data were used in the experiment and identifying traffic congestions were used as an example of clustering analysis. Results proved that the LRS-based clustering approach could dramatically decrease computational time in large data environment. Meanwhile, it could avoid the bias of DBSCAN that focuses on the number of GPS points rather than the number of cars. This could improve the accuracy of identified traffic congestions.

The experiment could be improved by using a set of real congestion records to train the cluster parameters, and thus to further validate performance of the LRS-based clustering approach. The proposed clustering approach could be extended to identify linear events.

Acknowledgments. This work is supported by China NSFC (41671387, 2018YFB2100704, and 41171348).

References

1. Ester, M., Keigel, H., Sander, J., Xu, X.: A density-based algorithm for discovering clusters in large spatial databases with noise. In: Proceedings of the Second International Conference on Knowledge Discovery and Data Mining (KDD) (1996)
2. Elnekave, S., Last, M., Maimon, O.: Incremental clustering of mobile objects. In: IEEE 23rd International Conference on Data Engineering, Istanbul (2007)
3. Lubr, S., Lazarescu, M.: Incremental clustering of dynamic data streams using connectivity based representative points. Data Knowl. Eng. **68**(1), 1–27 (2009)
4. Wang, Y., Lim, E.P., Hwang, S.Y.: Efficient mining of group patterns from user movement data. Data Knowl. Eng. **57**, 240–282 (2006)
5. Ma, S., Tang, S., Yang, D., Wang, T., Yang, C.: Incremental maintenance of discovered mobile user maximal moving sequential patterns. In: Lee, Y., Li, J., Whang, K.-Y., Lee, D. (eds.) DASFAA 2004. LNCS, vol. 2973, pp. 824–830. Springer, Heidelberg (2004). https://doi.org/10.1007/978-3-540-24571-1_72
6. Zhang, T., Ramakrishnan, R., Livny, M.: BRICH: an efficient data clustering method for very large database. In: International Conference on Management of Data (SIGMOD), New York, USA (1996)
7. Li, Z., Lee, J.-G., Li, X., Han, J.: Incremental clustering for trajectories. In: Kitagawa, H., Ishikawa, Y., Li, Q., Watanabe, C. (eds.) DASFAA 2010. LNCS, vol. 5982, pp. 32–46. Springer, Heidelberg (2010). https://doi.org/10.1007/978-3-642-12098-5_3
8. Nasraoui, O., Uribe, C.C., Coronel, C.R., Gonzalez, F.: TECNO-STREAMS: tracking evolving clusters in noisy data streams with a scalable immune system learning model. In: IEEE International Conference on Data Mining, Washington, USA (2003)
9. Park, N.H., Lee, W.S.: Cell trees: an adaptive synopsis structure for clustering multi-dimensional on-line data streams. Data Knowl. Eng. **63**(2), 528–549 (2007)
10. Choo, J., Jiamthapthaksin, R., Chen, C., Celepcikay, O., Giusti, C., Eick, C.: MOSAIC: a proximity graph approach for agglomerative clustering. In: Song, I.Y., Eder, J., Nguyen, T.M. (eds.) DaWaK 2007. LNCS, vol. 4654, pp. 231–240. Springer, Heidelberg (2007). https://doi.org/10.1007/978-3-540-74553-2_21
11. ESRI: ArcGIS Resource. rescources.argis.com/en/help/main/10.1/index.heml#/00390000000 1000000. Accessed 24 Aug 2015
12. OpenLR. https://www.openlr.org/introduction.html. Accessed 10 Oct 2015

A Method of Emergency Prediction Based on Spatiotemporal Context Time Series

Zilin Zhao[1], Zhiming Ding[1(✉)], and Yang Cao[2]

[1] Faculty of Information Technology, Beijing University of Technology, Beijing, China
zlzhao_bjut@qq.com, zmding@bjut.edu.cn
[2] School of Information, Beijing Wuzi University, Beijing, China
caoyangcwz@emails.bjut.edu.cn

Abstract. How to detect and predict the critical situation in large-scale activities is a very important research issue. The existing researches of emergency prediction are mainly focus on the micro events in some specific fields. Applying existing results directly to predict the critical situation in large-scale activity is a big challenge. In this paper, we show a novel method to predict emergency based on historical data analysis. We integrate relevant research results into a unified spatiotemporal model. Firstly, constructing the historical spatiotemporal context time series based on historical activity data. Then, dividing the time series into time period and time window. Finally, exploiting the time series' spatiotemporal patterns to predict the emergency of current activity. Experimental results show that the proposed method can achieve better prediction of large-scale activity emergencies in a specific venue.

Keywords: Spatiotemporal · Emergency · Time series · Event prediction

1 Introduction

Large-scale activities have been held more and more frequently in recent years. Once an incident occurs in an activity, it will cause heavy casualties and enormous economic losses. Therefore, how to detect and predict the critical situation in large-scale activities is a vital issue in both research and industry areas. Early prediction of emergencies can minimize losses and make a positive impact on emergency decision-making and resource scheduling. So, this research has great significance in the study of public safety issues.

Incidents have sporadic, highly systematic uncertainties, dynamics, chaotic effects [1] etc., which makes the traditional prediction of emergencies very difficult. In recent years, with the development of Information Technology, such as Cloud Computing, Big Data and Internet of Things, more technologies have been widely used in our society, such as video monitoring, face recognition, crowd density analysis, and navigation, etc. Large amount of spatiotemporal data has been generated. These data make a good foundation for Emergency Prediction (EP), and it is possible to achieve a better EP using big data analysis technology.

© Springer Nature Switzerland AG 2021
X. Meng et al. (Eds.): SpatialDI 2020, LNCS 12567, pp. 14–28, 2021.
https://doi.org/10.1007/978-3-030-69873-7_2

The spatiotemporal environment, personnel activities and other data of large-scale activities held by a specific venue have certain characteristics and regularity. Therefore, it is possible to predict whether an emergency will occur with the observation data accumulated by a specific venue historical activity. For example, the location, surrounding environments, internal structures, road characteristics and other spatial environment factors are basically unchanged for a long time. The schedules of the same type of activities (e.g., concerts or football matches, etc.) have some similarities (e.g., activity preparation, admission, progress, end and exit, etc.). The patterns of these data make good data mining analysis conditions for EP. In our study, we mine the internal correlation and the regularity within historical spatiotemporal data to get a better EP of the ongoing activity in a specific venue. Our main contributions are as follows:

1) Propose an activity spatiotemporal context model and give a method for similarity measuring in the context.
2) Show a method of historical spatiotemporal context time series construction.
3) Give Emergency Prediction algorithms based on historical spatiotemporal context time series.

The remaining part of this paper is organized as follows. Section 2 introduces relevant research. Section 3 gives the definition of spatiotemporal context model, discuss the construction method of historical Spatiotemporal context time series based on Spatiotemporal segmentation, and show the EP algorithm. Section 4 gives the experiments and its evaluation. Section 5 gives our conclusion.

2 Related Work

In Emergency Prediction area, Qiao et al. proposed a fuzzy evidential reasoning model [2]. Yang et al. showed a risk identification double-signal method for unconventional crisis based on immune danger theory [3]. Luo et al. given an event chain modeling method based on a three-layer frame [4]. Xu et al. showed a cross-media analytics method for emergency event detection and analysis [5]. The cross-media analysis provides a good idea for information fusion. In our work, we combine both physical and media data for large-scale EP prediction.

Historical data analysis methods often find regularities that conventional methods could not be [1]. Time series will show better characteristics trends within historical data, and it is an effective tool for spatiotemporal data analysis. In time series study, Lexiang Ye et al. first give time series shapelets [6]. Ji et al. presented a fast shapelet discovery algorithm based on important data points [7]. Chen et al. presented a non-equal time series similarity measure algorithm [8]. Xu et al. propose a data service request prediction approach based on historical user behavior time series analysis [9].

Existing EP based on historical data is mainly targeted at specific fields, aspects, and micro-events, etc., such as crowd density prediction [10], road prediction [11], crowd classification [12, 13], emotional prediction [14], abnormal behavior prediction [15], etc.

The output of these research results mostly are attribute tags (e.g., the crowd identification classification output is represented as a classification attribute), and each result is generally a classification for the completion of a specific target task. These results can be applied separately to monitoring and tracking of large-scale activities. E.g., crowd density monitor can be deployed at the entrances and exits of activity venue, aisles and seats. Two monitors in different location have different crowd density within the same activity stage. If monitoring devices identify a specific individual (such as a suspect), the activity area where the individual appeared will have a higher risk of emergency. Tracking Weibo, WeChat, Headlines and other hot topics can also help judge emergencies risks. Each independent research result can make emergency predication from its special viewpoint. But it is difficult to predict entire activity emergency because multi viewpoints must be considered together. In this paper, we creatively propose a method which integrates these results into a unified spatiotemporal context model, and does EP based on historical spatiotemporal context data.

3 Our Model

3.1 Spatiotemporal Context Model

Definition 1 (Spatiotemporal Context Model -STCM). Spatiotemporal Context Model is a 6-tuple:

$M = <G, S, T, H, f^w, f^R>$, in which:

- $G = \{G_0, G_2, \ldots, G_s\}$, G is the data set of previous large-scale activities in a place. Each G_i is an activity.
- $S = \bigcup_{i=0}^{k} S_i$, $S_i = \bigcup_{j=0}^{m} x_{ij}$, S is the monitor item set of G. Each monitor item S_i consists of a series of monitor attributes x_{ij}. The monitor items are the researches result deployed and the real monitoring device.
- $G_i^{th} = \{S_1^t, S_2^t, \ldots S_k^t\}$
  $= \{\{x_{11}^t, x_{12}^t, \ldots x_{1m}^t\}, \ldots \{x_{k1}^t, x_{k2}^t, \ldots, x_{km}^t\}\}$
  G_i^{th} is the spatiotemporal context of G_i at time t and region h. T is the set of observation time, $t \in T$; H is the set of regions within the activity site, $h \in H$.
- f^w and f^R are the functions on G. Given activity g_0, let t_0 be the current moment, and h_0 be the activities monitoring region. $f^w(g_0, t_0, h_0)$ returns the time window which the activity g_0 located in the spatiotemporal segmentation.

The problem in this paper can be defined as:

$$f^R(G, g_0, t_0) \rightarrow p, p \in [0, 1]$$

How to get function f^R is the goal of this paper.

Assume that different monitor items have different monitor attributes, G_i^{th} can be simplified into the following Eq. (1), in which, $val(x, t, h)$ is the value of monitor attributes x at time t and region h.

$$G_i^{th} = \left\{x_1^{th}, x_2^{th}, \ldots, x_n^{th}\right\} = \bigcup_{x \in Gi} val(x, t, h) \tag{1}$$

The spatiotemporal context model can well express the historical spatiotemporal data space G (see Eq. (2)).

$$G = \bigcup_{i=0}^{s} \bigcup_{t \in T} \bigcup_{h \in H} G_i^{th} = \bigcup_{i=0}^{s} \bigcup_{t \in T} \bigcup_{h \in H} \bigcup_{x \in Gi} val(x, t, h) \qquad (2)$$

Given an activity G_i, we can rearrange the data in G_i by the data generate time. As shown in Fig. 1, a node means a monitor attribute value.

Then we can get a time series G. We called it Original Spatiotemporal Context Time Series (OCTS).

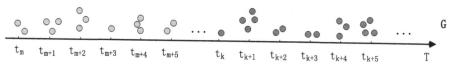

Fig. 1. Spatiotemporal Context Time Series

3.2 Time Series of Historical Spatiotemporal Context Based on Spatiotemporal Segmentation

In real activity scenario, data in different time periods or spatial regions have different data characteristics. In terms of time, the monitor attribute value is periodic, (For example, in the activity preparation, admission, progress, end, exit and other stages of reciprocating. Concert ring environment and atmosphere with the singer's songs start-stop periodic changes, etc.). The characteristics and patterns of activities' monitor attribute value in the same period are usually similar. In terms of space, active physical space usually contains different regions. (Such as external plaza, parking lot, road, internal grandstand A, B, rostrum, stage, lounge, etc.). Due to the different tasks, functions and field personnel in different regions, the monitor attribute value of different spatial regions also has different characteristics.

In order to observe the patterns of monitoring data better, we rearrange the historical spatiotemporal data according to two dimensions of time and space.

1) For the time dimension, we segment the data space into Time Periods and Time Windows according to the active time.
2) For the spatial dimension, we segment the data space into Grids according to the physical position. The specific method is described as follows.

Spatiotemporal Segmentation
Time Segmentation. Considering the real situation of the activities, the process of activities can be divided into different periods. So, we introduce the Time Period and Time Window.

Time Period (TP) Is the time interval in which the activity cycle changes, such as a week or a day. It used to express the segmentation of activities according to time. Time Window (TW) is the further segment of the Time Period, such as hour, half an hour and one minute. Time window is used to express the detailed time segmentation within a period.

We divide each activity in historical spatiotemporal data space G using TP and TW. Let p be the TP length, and w be the TW length. Then the observation time axis T is segmented into several finite periods of length p. As shown in Fig. 2 (a), each activity has its own start time before segmenting the time period. The activity data is arranged along the time axis in order of the time it collected (we assume that only one activity can be held in the venue at the same time). After segmenting TP, all activities are mapped to a unified start time (Fig. 2(b)). Now, the time of the monitor attribute value is converted to the relative time *relt* to start time of its TP. Monitor attribute values of each activity are transformed into a subsequence of OCTS with TP and TW. In Fig. 2(b), e.g., all the time periods *tp* are segmented into 3 time-window tw_{1-3}. Three windows can express the stages of activity admission, activity development and exit, respectively. For tw_1, the personnel flow of entrance and exit may often increase rapidly, and then it tended to stabilize. Ultimately, the rate gradually decreases and tends to zero. The crowd density of venues gradually increases and tends to constant, and so on. For tw_2, the monitor attribute value pattern may be different. It may present that the flow of entrance or exit extremely low and remains substantially constant. The density of personnel in the venue is high and basically constant, and so on. Therefore, the time series data set integrated by tw_1 has the time characteristics of activity admission.

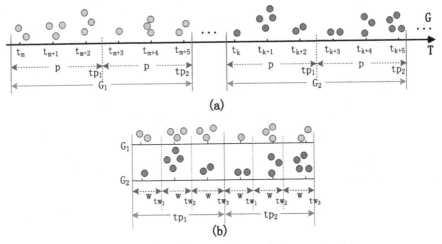

Fig. 2. Spatiotemporal Context Time Series

After transformation, we can compare the monitor attribute value collected from different activities under the same condition. And it is convenient for us to summarize the patterns of historical activities in different TP and TW.

Space Segmentation. In real activity monitoring, each monitor item can only focus on a specific region. For example, camera h monitors region A, each monitor item's output describes the crowd density in region A. It is reasonable to have a high crowd density at the entrance and a low crowd density in the grandstand area during the entry stage. However, during the activity development phase, the low crowd density at the entrance and the high crowd density in the grandstand is normal. At this stage, if the monitor attribute value showing the patterns of the active entry stage may be abnormal. In order to discover and express the different regions' pattern of monitor attribute value scientifically and accurately, we use Grid (GR) to segment the OCTS according to the physical region. Furthermore, the activity monitoring data not only covers the geography space include seats, roads inside and outside the venue, but also covers the social media space such as Weibo. Network space data often haven't local location characteristics, so it does not need space segmentation.

Space segmentation solves the problem that the output of the same monitor attributes has different standards in different regions. Moreover, in the same region, different monitor attributes have different importance to the evaluation result of EP. So, we assign different weights to each monitor attribute and adjust the accuracy of EP by adjusting the weights of attributes.

Let $gr = \bigcup_{l=0}^{u} gr_l$ be the grid Set. For $\forall gr_l$, the attribute weight set is expressed as:

$$W = \{w_1, w_2, \ldots, w_n\} = \bigcup_{x \in Gi} w(x, gr_l) \tag{3}$$

Then, the weight set of OCTS is shown below:

$$W = \bigcup_{gr \in H} W^{gr} = \bigcup_{gr \in H} \bigcup_{x \in Gi} w(x, gr_l) \tag{4}$$

After space segmentation, the OCTS is divided into subsequences of multiple regions. Moreover, there are multiple active time subsequences for any one region. So, the Historical Spatiotemporal Context Time Series (HSTCTS) is a collection of time series with time and space segmentation.

Construct HSTCTS Base on Spatiotemporal Segmentation

Algorithm 1. HSTCTS Construction Algorithm.

Input: *ots*: OCTS; *p*: the length of TP; *w*: the length of TW; *glen*: the size of GR;

Output: *hts*: HSTCTS;

1) *Rank (ots,G_i^{th} ∈ ots);* // Sort the monitor attribute value G_i^{th} in OCTS by time *t*
2) **for each** *sp* ∈ *hts* **do** // Initialize the HSTCTS
3)   *sp* ← *Ø;* // *sp* is a time subsequence with time and space segmentation
4) **end for**
5) **for each** G_i^{th} ∈ *G* **do** // Calculate the window number *twi* and grid number *gri* of G_i^{th}
6)   *relt* ← *t mod p;* //*relt* is the relative time relative to the start time of its TP
7)   *twi* ← *relt mod w;*
8)   *gri*← *h /* glen;
9)   *sp(tw, gr)* ← *sp(tw, gr)* ∪ { G_i^{th} }; //Add to the corresponding time subsequence *sp* of HSTCTS
10) **end for**
11) **return** *hts;*

3.3 Emergency Prediction Based on HSTCTS

The method of prediction the current activity emergencies based on HSTCTS can be divided into two steps as follows.

Step1: In HSTCTS, find a collection of historical time windows similar to the window in which the current activity monitor attribute value located.

Step2: Predict whether emergencies will occur in the next window according to the occurrence of the historical emergencies in similar time windows.

In the following, we will elaborate on the prediction method of the emergency.

The Similarity Measurement of Historical Spatiotemporal Context. It can be seen from the definition 1 that, in essence, querying similar spatiotemporal contexts is a retrieval process of monitor attribute value in high-dimensional space. Let the activity data set G be the historical spatiotemporal data space. Then, the monitor attribute value of spatiotemporal context $G_i{}^{th}$ can be represented as a vector: $G_i{}^{th} = <x_1{}^{th}, x_2{}^{th}, ...,$ $x_n{}^{th}>$ ($G_i \in G, t \in TW, h \in GR$).

Definition 2 (Similar Spatiotemporal Context): Given an arbitrarily small positive number ε, if $\exists G_i^{t1}, G_j^{t2}, t_1, t_2 \in tw$ can meet the condition $dist(G_i^{t1}, G_j^{t2}) \leq \varepsilon$, we called G_i^{t1} and G_j^{t2} are the similar spatiotemporal context, and denote them as $G_i^{t1} \sim G_j^{t2}$. When $\varepsilon \to 0$, the spatiotemporal contexts G_i^{t1} and G_j^{t2} tend to be the same, but the same is not transitive.

We use Euclidean distance as the distance measure of the two monitors attribute value in the same time window. For $\forall G_i^{t1}, G_j^{t2}, t_1, t_2 \in tw$, the distance of monitor attribute value G_i^{t1} and G_j^{t2} is shown below:

$$dist\left(G_i^{t1}, G_j^{t2}\right) = \sqrt{\left(\sum_{x \in D} val(x, t_1, h) \times w(x, GR) - val(x, t_2, h) \times w(x, GR)\right)^2} \tag{5}$$

Emergency Prediction Algorithm Based on HSTCTS. In real activity scenario, the patterns of monitor attribute value may have certain similarities after spatiotemporal segmentation. If most of the similar historical windows which the monitor attribute value belongs to has occurred the same type of emergency, this type of emergency will likely occur in the future period in current activity.

We design an Emergency Prediction algorithm based on HSTCTS (Algorithm 2). First, acquiring all the historical monitor attribute value, which is the same spatiotemporal segmentation as current monitor attribute value from HSTCTS. Secondly, according to definition 2, we calculate the similarity between the current monitor attribute value and the historical monitor attribute value using k-Nearest Neighbor. Recording the K maximum similarity window where the historical monitor attribute located, and collecting them in the set *topK*. At this time, we can get the type of windows *Maxtype* that appear most. Eventually, we make statistics on collection *topK*. Let the grid location be the incident occurrence region. Calculating the ratio of maxtype windows' number and similar time windows' number $|topK|$. Let this ratio be the probability of the *Maxtype* event occurring in a future time period. If the probability exceeds 50%, the type of the next window of the current period is determined as this event type; otherwise, it is determined as the common event type. Next, we give the pseudo code of the Emergency Prediction algorithm.

Algorithm 2. Emergency Prediction algorithm based on HSTCTS.

Input: *hts*: HSTCTS; G_g^{th}: the monitor attribute value of current activity g; *p*: the length of TP; *w*: the length of TW; *glen*: the size of GR;

Output: *R*: the result set of prediction (including location, event type and probability);

1) Let *type* be the event type of current activity, and initialize to the common event type; MaxType is the type of events that occur most frequently in similar Windows
2) Calculate the window number *twi* and grid number *gri* of G_g^{th}
3) *sp* ← *getSp (twi, gri, hts)*; //The function *getSp* find the time subsequence *sp* in *hts* with *twi* and *gri*
4) **for each** $G_i^{t'h} \in sp(twi, gri)$ **do**
5)   **if** $G_i^{t'h} \sim G_g^{th}$ **then**
6) $tw \leftarrow f^w (G_i, t',h)$;
7)   **if** $tw \cap topK = \emptyset$ **then**
8) $topK \leftarrow topK \cup tw$;
9)   **end if**
10)   **end if**
11) **end for**
12) **for each** $tw \in topK$ **do**   //Update *Maxtype*
13)   *updateMaxtypeType (Maxtype, $f^t(tw)$)*;
         //$f^t (tw)$ is the judgment function of the event type for the time window *tw*
14) **end for**
15) **if** $type/|topK| > 0.5$ **then** // Statistics the *topK* collection, and given the *type*
16) $type \leftarrow MaxType$;
17) **end if**
18) $R \leftarrow (type/|topK|, type, gr)$;
19) **return** R;

---

4   Experimental and Result Analysis

4.1   Experiment Setup

We have developed a simulation program to simulate a large number of historical activity data using Java Language. The development environment is JDK_10 and Eclipse. The operating system is Windows 10. Our machine has 12 GB RAM with an Intel Core i5-4200M CPU(2.50 GHz).

The emergencies in each space segmentation directly determines the emergencies in the whole activity. Therefore, we only simulate one spatial segmentation as an example to analyze the experimental results. For simplicity, the monitor attributes weights of the spatial segmentation are all 1.

According to algorithm 1 and 2, the scheduler simulated the construction of historical spatiotemporal data space (construct the OCTS), spatiotemporal segmentation (construct the HSTCTS) and current activity monitor attribute value. The experiment was divided into three rounds marked as A, B and C. Each round has 100 times of historical activities. Each activity is divided into 3 TP (simulating the event for 3 days). Twelve time windows per TP (2 h is a time window). 60 monitoring data per TW (a data sampling in 2 min). Each data is described by 20 monitor attributes.

To simulate the actual activity patterns better, we divide the anomalies into three types, which are Monitor Attribute Anomaly, Data Anomaly and The Window Anomaly. The window anomaly is used to simulate the abnormal activity in a certain period of time. The data anomaly express the abnormal monitoring data collected at one time. Monitoring value anomaly simulates an anomaly in the output of a monitoring device or item. we describe the abnormal standard as follows:

1) Monitor Attribute Anomaly: The attribute value within the range of the abnormal value.
2) Data Anomaly: Half and more than half of the monitor attributes anomaly.
3) The Window Anomaly: Half and more than half of the data anomaly.

The emergencies are rare in real activity scenario, and the anomalous data may very sparse in the real data space. If we simulate the abnormal data strictly according to the real activity scene entirely, the data volume will be very huge, and the simulation will have low efficiency. In our experiment, for better efficiency and easy comparison of results, we set one third of the history windows be abnormal event type. Generate more abnormal data will help us to test the effectiveness of the algorithm.

There are many kinds of emergencies in real activity scenario. We just simulated five event types. The type 0 is normal event type, and type 1 to type 4 are emergencies. Two third of the history windows are normal event type. One third are abnormal event types random in type 1 to type 4. If a window is abnormal, in this window, the number of abnormal data is random in the range [30,60). Otherwise, it random in the range [0,30). For abnormal data, the number of abnormal attribute random in the range [10,20). Otherwise, it random in the range [0,10). Values of all monitor attributes are randomly generated within the ranges, which are defined separately within different round.

The simulation program randomly generates current activity data. Each data in this time window is manually marked according to the data anomaly standard. This data is all the data of an abnormal time window (60 in total).

4.2 Experimental Results

As mentioned above in Sect. 3.2, in real activity scenario, the range of monitor attribute values for normal and abnormal data shows a regular change over time. Therefore, we simulate different patterns of the window by giving different normal and abnormal ranges. In Round A, we generated data space and current data completely randomly to observe the effectiveness of the prediction algorithm under irregular conditions. Round

B and C respectively simulated the data space and the current data with weak and strong patterns. The aim is to obtain the influence of the patterns' strength on the prediction accuracy. In the A, B and C rounds, the comparison between historical space and window space can give the effect of time period and window segmentation on prediction accuracy.

Table 1 to 3 show the results of 3 rounds. In each round, ten groups of current data were predicted in the same historical space. Moreover, each group can get 60 predictions. Comparing the manual data markers with the detected results and calculating the accuracy (accuracy = the number of correct/60). Each row in three tables is a complete result.

Results of Round A. In Round A, all monitor attribute values in the historical space and current activity data are generated completely randomly in the range [0,100). For the monitor attribute values (20 in total), The abnormal value range is [50,100), and the normal value range is [0,50).

Table 1. Results of Round A

HSCN	HSA	WSCN	WSA
44	73.33%	28	46.67%
41	68.33%	22	36.67%
41	68.33%	57	95.00%
41	68.33%	57	95.00%
48	80.00%	28	46.67%
43	71.67%	29	48.33%
37	61.67%	14	23.33%
50	83.33%	19	31.67%
51	85.00%	54	90.00%
43	71.67%	41	68.33%

In Table 1, 2, and 3, HSCN is Historical Space Correct Number. HSA is Historical Space Accuracy. WSCN is Window Space Correct Number. WSA is Window Space Accuracy.

Results of Round B. In Round B, we randomly divide 12 time windows in the historical space into four groups. As for the same monitor attribute under the same group window, the normal and abnormal value range is the same. Current activity data's patterns follow the corresponding time windows' patterns in historical space. The attribute patterns are simulated as follows:

1) Dividing the attributes value range [0,100) into [0,20), [20,40), [40,60), [60,80) and [80,100) five sub-ranges, in order to express the different normal or abnormal value ranges.

2) Each attribute (20 in total) randomly selects a range from the five sub-ranges as the abnormal value range. The normal value range is any region in the interval [0,100] except the abnormal value range.

Table 2. Results of Round B

HSCN	HSA	WSCN	WSA
42	70.00%	52	86.67%
42	70.00%	57	95.00%
49	81.67%	50	83.33%
44	73.33%	31	51.67%
45	75.00%	46	76.67%
47	78.33%	54	90.00%
40	66.67%	52	86.67%
44	73.33%	29	48.33%
39	65.00%	37	61.67%
24	40.00%	58	96.67%

Results of Round C. In Round C, the simulation randomly generated different normal or abnormal patterns for 12 kinds of time windows. Therefore, for the same monitor attribute with the same window number, the normal and abnormal value range are the same. The monitor attribute of current data meets the corresponding window data patterns in historical space. Furthermore, the attribute patterns are simulated as Round B.

4.3 Results Discussion

From Fig. 3, we can see that when monitor values are completely random in historical and current space, results in historical space accuracy is mostly better than the split window space. That is because the historical space has much larger data volume than the window space. The large amount of emergency data makes the prediction more comprehensive. Although the window space accuracy is lower than the historical space, its maximum can be up to 95%. It shows that the window space can realize EP even under completely random conditions. Moreover, because the amount of data used for measurement is greatly reduced, the calculation speed in the window space is much faster than the historical space. From this aspect, it is much better than the historical space. It can better meet the needs of real-time EP (Fig. 4 and Fig. 5).

Table 3. Results of Round C

HSCN	HSA	WSCN	WSA
41	68.33%	48	80.00%
28	46.67%	27	45.00%
39	65.00%	52	86.67%
38	63.33%	50	83.33%
23	38.33%	19	31.67%
40	66.67%	55	91.67%
37	61.67%	52	86.67%
44	73.33%	55	91.67%
45	75.00%	60	100.00%
32	53.33%	51	85.00%

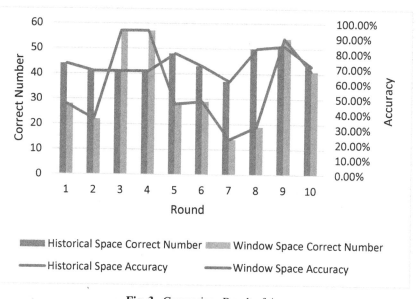

Fig. 3. Comparison Result of A

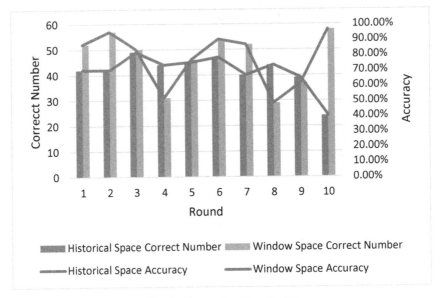

Fig. 4. Comparison Result of B

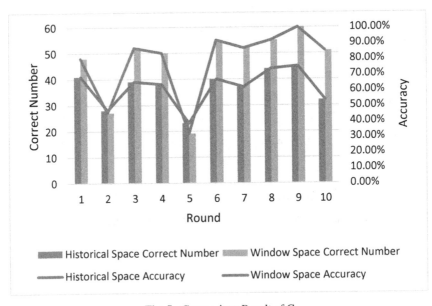

Fig. 5. Comparison Result of C

5   Conclusion

In this paper, we propose the Spatiotemporal Context Model, a novel method which integrates relevant research results into a unified spatiotemporal model. The Spatiotemporal Context Model is able to predict emergencies of activities by spatiotemporal segmentation and data mining analysis. The experimental results show that the proposed method is sensitive to the data patterns. And our model can achieve real-time prediction of large-scale activity emergencies in specific venues. In our future work, we will make a further deep study of relevant issues, such as optimize the event type discrimination, analysis the correlation between different physical regions and optimize the assignment of attributes weight. As for event type discrimination, different types emergency has different warning level. Different level will make different action. The relationship of Type-Level-Action should be an interesting problem that we will make further study.

Acknowledgment. The work was partially supported by the National Key R&D Program of China under grant number 2017YFC0803300, National Natural Science Foundation of China under grant number 91546111 and 91646201, Beijing Municipal Education Commission Science and Technology Program under grant number KZ201610005009 and KM201610005022.

References

1. Wang, G.Q., Liu, Y., Yang, P., et al.: Study on decision-making method for complex crisis. Syst. Eng. Theory Pract. **35**(10), 2449–2458 (2015)
2. Qiao, X.J., Li, Y.J., Chang, B., et al.: Modeling on risk analysis of emergency based on fuzzy evidential reasoning. Syst. Eng. Theory Pract. **35**(10), 2657–2666 (2015)
3. Yang, Q., Liu, X.X., Yang, F., et al.: Risk identification double-signal method for unconventional crisis based on immune danger theory. Syst. Eng. Theory Pract. **35**(10), 2667–2674 (2015)
4. Luo, P.Z., Chen, X.: Research on emergency event chain modeling and retrieval. J. Huazhong Univ. Sci. Technol. Nat. Sci. Edn. **S1**, 440–444 (2015)
5. Xu, W., Liu, L.Y., Wang, M.M.: Emergency event detection and analysis for emergency management based on cross-media analytics. Syst. Eng. Theory Pract. **35**(10), 2550–2556 (2015)
6. Ye, L., Keogh, E.: Time series shapelets: a new primitive for data mining. In: Proceedings of the 15th ACM SIGKDD International Conference on Knowledge Discovery and Data Mining, pp. 947–956. ACM (2009)
7. Ji, C., Zhao, C., Pan, L., et al.: A fast shapelet discovery algorithm based on important data points. Int. J. Web Serv. Res. (IJWSR) **14**(2), 67–80 (2017)
8. Chen, H.L., Gao, X.D.: Research on wavelet-based time series similarity measure and cluster. Stat. Decis. **35**(11), 17–22 (2019)
9. Xu, J., Du, L., Song, C., et al.: Big data service request prediction based on historical behavior time series. In: Proceedings of the 2nd International Conference on Big Data Technologies, pp. 77–81. ACM (2019)
10. Sindagi, V.A., Patel, V.M.: CNN-based cascaded multi-task learning of high-level prior and density estimation for crowd counting. In: 2017 14th IEEE International Conference on Advanced Video and Signal Based Surveillance (AVSS), pp. 1–6. IEEE, (2017)

11. Ding, Z., Ren, F., Cai, Z., et al.: Urban road congestion condition discrimination research based on vector features. In: IEEE Third International Conference on Data Science in Cyberspace (DSC), pp. 92–98 (2018)
12. Li, Q., Zhao, X., He, R., et al.: Recurrent prediction with spatio-temporal attention for crowd attribute recognition. IEEE Trans. Circ. Syst. Video Technol. (2019)
13. Cao, Y., Si, Y., Cai, Z., et al.: Mining spatio-temporal semantic trajectory for groups identification. In: 2018 IEEE 9th Annual Information Technology, Electronics and Mobile Communication Conference (IEMCON), pp. 308–313. IEEE (2018)
14. Patwardhan, A.: Edge based grid super-imposition for crowd emotion recognition. arXiv preprint arXiv:1610.05566 (2016)
15. Selvi, S.S.: Crowd recognition system based on optical flow along with SVM classifier. Int. J. Electric. Comput. Eng. **9** (2019). 2088–8708

A Real-Time Driving Destination Prediction Model Based on Historical Travel Patterns and Current Driving Status

Jintian Wang[1] , Zhipeng Gui[1,2,3(✉)] , Yunzeng Sun[1] , Huayi Wu[2], and Zifan Yu[1]

[1] School of Remote Sensing and Information Engineering, Wuhan University, Wuhan, China
zhipeng.gui@whu.edu.cn
[2] State Key Laboratory of Information Engineering in Surveying,
Mapping and Remote Sensing, Wuhan University, Wuhan, China
[3] Collaborative Innovation Center of Geospatial Technology, Wuhan, China

Abstract. As an important application of location-based services (LBS), driving destination prediction support route planning, service recommendation and vehicle scheduling, etc. However, destination prediction in a real-time manner is challengeable due to the influence of both historical travel patterns and current driving status, which need to be considered in modeling. To fill this gap, we proposed a real-time prediction model with two modules, i.e., hierarchical temporal attention module and status-aware prediction module, which utilize driver's historical Original-Destination (OD) sequences and current travel trajectories as inputs respectively. More specifically, the hierarchical temporal attention module can effectively process the OD sequences under the calendar period. The status-aware prediction module achieves the prediction according to the current travel status and the key travel location identification in current trajectory. Comparative experiments with baseline models verified the validity of our model. Further analyses discussed the factors that affect the prediction performance from the perspective of distance, time span and grid partition granularity.

Keywords: Attention mechanism · Long Short-Term Memory (LSTM) · Current travel trajectory · Historical origin-destination sequences · Driving regularity and uncertainty

1 Introduction

Location prediction has widespread applications in many fields, such as vehicle scheduling, advertising, and emergency response of public events. With the improvement of data collection capability, massive personal driving trajectories can be recorded, which provides strong support for learning individual-level moving patterns, and makes driving destination prediction available. However, providing a high-precision prediction in a real-time manner remains great challenging.

Traditional models are commonly used historical trajectories as the basis to predict destination. For example, the Markov model, which is widely used in location prediction,

X. Meng et al. (Eds.): SpatialDI 2020, LNCS 12567, pp. 29–43, 2021.
https://doi.org/10.1007/978-3-030-69873-7_3

learns the state transition probability between representation features in historical trajectories for mining travel patterns. However, these probabilistic models only depend on the states in several previous steps, which are not suitable for long-term dependency mining in sequence data. In deep learning models, Recurrent Neural Networks (RNNs) are commonly used to process data with time sequences, which can express the dependencies between sequential trajectories better than probabilistic models. However, RNNs also need to integrate other information to strengthen their ability of expressing travel characteristics.

To address above problems, this paper proposes a real-time driving destination prediction model based on the historical travel patterns and current driving status (HTPCDS). The remainder of the paper is structured as follows. Section 2 reviews related works. In Sect. 3, the paper outlines the model framework, and explains the hierarchical attention mechanism module and the real-time prediction module in detail. Section 4 validates the effectiveness of the model through experiments and further discusses the factors affecting the prediction performance of the model. Section 5 draws conclusions and discusses future research.

2   Related Work

2.1   Prediction Using Current Traveling Context

The traveling context embodies rich semantic that reflects human activities. For example, road network structures and urban functionalities of POI [1] restrains driving behavior. Meanwhile demographic characteristics and social media footprints [2] of an individual may reflect driving intentions and social interactions. However, most of these semantics need to be obtained by linking heterogeneous data sources and conducting sophisticated data mining, which makes it hard to be used in travel destination prediction scenarios. In contrast, the driving status embedded in trajectory data is also helpful to infer destination but rarely used in driving prediction. Driving status such as stopping, turning or accelerating may reveal the travel intentions of a driver. For instance, when the vehicle turns, the speed will be slowed down and the azimuth of vehicle will change, which can reflect current traveling context in driving. So, we can utilize locations with critical driving status in the trajectories to strengthen the prediction ability.

Although the current driving status can explain the behavior of driver, it is difficult to be utilized by prediction models for mining the implicit associations. The attention mechanism provides an effective way to solve the combination of driving status and trajectories, which is initially used in text analysis [3]. It can generate traveling context vectors by weighing the similarity of adjacent track locations. Hence, we constructed a status-aware prediction module to extract spatial attention by utilizing current travel context including speed, azimuth, and acceleration [12].

2.2   Mining of Historical Travel Patterns

The historical trajectories record the frequent driving routes and travel habits of an individual for mining spatiotemporal driving patterns. Markov decision process [4] is often used to location prediction according to the transition probability from the current location to the next location. However, it ignores the continuity of the sequence and only considers the impact of past few historical locations. The location prediction based on historical trajectories can also use decision tree [5], which build all possible ways within the origin-destination pairs but neglect the relationships between each attribute. The above methods focus on the spatial dimension characteristics of historical trajectories, but temporal feature mining is insufficient. The neural networks such as Long Short-Term Memory (LSTM) are excellent in the processing of sequential sequences in time dimension because it can remember and update sequence information. LSTM has been used to predict travel destination. However, historical travel patterns are rarely considered because the patterns in the calendar period are hard to be mined without hierarchical structures [6]. For instance, the travel trajectories in last weekend may have more influence on the travel intentions of this weekend rather than this Friday. Hence, beside utilizing driving status, we also adopt stacked attention layers on basis of LSTM to extract travel regularities in different temporal levels through long-term historical OD sequences.

2.3 Temporal Attention Mechanism

In location predict models, the attention mechanism is commonly used to extract high relevant trajectories between adjacent trajectories in spatiotemporal dimension. The attention mechanism has been used in the encoder-decoder to improve the language translation [7] and location prediction [8, 9].

Attention mechanism has many variants including self-attention, global attention, and local attention [10, 11], etc. Self-attention mechanism focuses on the correlation of internal enter and outputs an attention vector. Global attention with a broad perspective extracts features of all input sequences. In contrast, local attention focuses on inner connections between contiguous inputs. However, they are inefficient when dealing with multi-level time sequences and needed to be integrated. Hence, we adopt a hierarchical attention mechanism which stacks attention layers to extract different-level features in calendar periods. By using hierarchical attention mechanism, we can effectively extract daily and weekly driving patterns.

Therefore, this paper proposes a real-time destination prediction model for driving trajectories by combining the historical travel patterns and current driving status together. We used LSTM and temporal attention to capture the impact of historical travel patterns on travel. Meanwhile, we used current driving status, including speed, distance and steering angle, to mine real-time travel intentions.

3 Methodology

3.1 Model Framework

As shown in Fig. 1, The proposed model includes three modules, i.e., input module, hierarchical temporal attention module and status-aware prediction module. Input module preprocesses the current trajectories and historical sequences. Hierarchical temporal attention module takes historical OD sequences as input for mining different-temporal-level traveling patterns in calendar period. While, the status-aware prediction module learns the spatial patterns from current trajectories and outputs the destination prediction by considering both historical and current attentions.

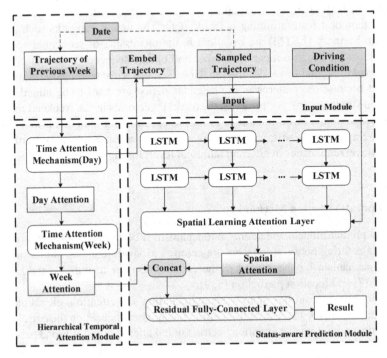

Fig. 1. The framework of the proposed prediction model.

To associate the hierarchical temporal attention module with the status-aware prediction module, the historical OD sequences are encoded into fixed length vectors and transferred to the input module. The input module combines current trajectories, semantic features and historical encoding vectors to generate the input of the status-aware prediction module. Meanwhile, the output of hierarchical temporal attention module is spliced with the output of spatial attentions, and transfers the splicing result into the residual fully connected layer to output the destination prediction. In addition, we use the occupancy grid map (OGM) [6] to divide the historical OD sequences and the current trajectories into grids, which strengthens the spatial correlation between adjacent points.

3.2 Hierarchical Temporal Attention Module

The structure of the hierarchical temporal attention module [11] is illustrated in Fig. 2. More specifically, it contains two sub processes including the local attention to find the patterns of day in historical sequences and the global attention to extract more significant weekly features. Day attention focuses on OD location relationships in one day, which reflects the driver's daily transference but has inadequate capacity to learn the patterns in long-term traveling records. To conclude the general patterns, Week attention further extracts sequential relationships on weekly basis according to the day attentions, which can reflect the importance of daily travel throughout the week.

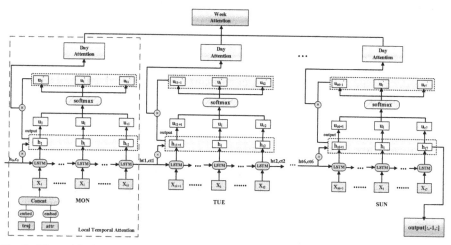

Fig. 2. Hierarchical time attention mechanism module. Where *traj* refers to historical grid sequences, X_i refers to the OD point, *attr* refers to the input feature value, *output* takes the value of the last time step of the hidden layer, which represents the equal-length vector after sequence encoding.

Local Temporal Attention. The local temporal attention mechanism focuses on the daily travel trajectories of a driver, which allocates different attentions to daily features according to the degree of correlation with the target location. To extract daily travel patterns, we generated daily attention vectors according to Eq. 1.

$$
\begin{aligned}
u_i &= V_a^T tanh\big(W_a'\,[h_{(T-1)}, c_{(T-1)}] + W_a h_i + b_a\big) \\
u_i' &= softmax(u_i) = (exp(u_i))/(\textstyle\sum_{(i=1)}^t u_i) \\
H_T &= \textstyle\sum_{i=1}^t u_i h_i
\end{aligned}
\tag{1}
$$

Where h_{T-1}, c_{T-1} are the hidden state finally output the day before, h_i is the hidden layer state at the i-th time step, u is the weight, H_T denotes the vector of the final output of the day.

Moreover, in order to simulate the travel patterns for a week, the local temporal attention mechanism has been designed to consist of seven modules with the same structure arranged in sequence. The status of the previous day is transferred to the next day, so that context information is formed between adjacent modules.

Global Temporal Attention. After the historical sequences are processed by the local time attention mechanism, attention vectors containing features of each day $\{H_1, H_2, H_3, H_4, H_5, H_6, H_7\}$ has been generated. To mine more representative travel patterns, we further propose a global time attention mechanism on basis of local attention, which increases the proportion of days with higher correlation with the current prediction in the weight distribution, and outputs attention vectors that conform to weekly patterns. The formula is shown in Eq. 2:

$$w_i = V_a^T tanh\left(W_a' [h_{(T-1)}, c_{(T-1)}] + W_a h_i + b_a\right)$$
$$w_i' = softmax(w_i) = (exp(w_i))/(\textstyle\sum_{(i=1)}^{7} w_i)$$
$$H = \textstyle\sum_{i=1}^{7} w_i H_i$$

(2)

In Eq. 2, the feature vector of each day is combined with the state information of the previous day to calculate the weight. After normalization, the global temporal attention vector is output in the form of a weighted sum.

3.3 Status-Aware Prediction Module

The status-aware prediction module receives the current trajectories and outputs a geographic coordinate as the final predicted destination result. It composes of two sub-modules, including the spatial learning sub-module and the location prediction module.

Spatial Learning Sub-module. The sub-module shown in Fig. 3 is used to capture the correlation of the current trajectory sequences in the spatial dimensions by using scorer layer and trajectory spatial attention layer.

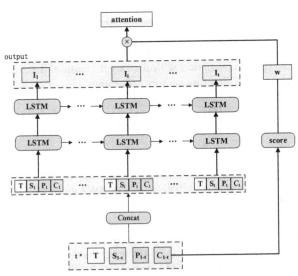

Fig. 3. The spatial learning sub-module. Where T is time, S is the semantics of points, P refers to sub-trajectory, C represents real-time driving status, w is the weight calculated by the score layer.

The sub-module adopts two stacked LSTM layers because of its better performance than single-layer LSTM. The scorer layer is used to output current attention vectors which captures the spatial location information related to the current driving status and road network structures by analyzing real-time motion characteristics including vehicle speed, distance, and steering angle. The distance is calculated based on adjacent trajectory points according to Eq. 3.

$$dis_i = \sum_{n=1}^{i} Dis\left(P_{(n-1)}, P_n\right)$$
$$\Delta Dis(P_{n-1}, P_n) = h(d/R) = h(\varphi_2 - \varphi_1) + \cos(\varphi_1)\cos(\varphi_2)h(\Delta\lambda) \qquad (3)$$
$$h(\theta) = \sin^2(\theta/2) = (1 - \cos(\theta))/2$$

where d is the distance between two points on the sphere, R is the radius of the sphere, θ represents the angle radian between the center of the adjacent track points.

According to the Eq. 4, the driving speed is expressed by the average speed of the current point and the previous point.

$$spd_i = \frac{\Delta Dis(P_{i-1}, P_i)}{\Delta time(P_{i-1}, P_i)} \qquad (4)$$

The vehicle steering angle ste_i is calculated based on the angle between the current point P_i and two adjacent points P_{i-1}, P_{i+1}. The three-driving status mentioned above can be used to evaluated current location importance according to Eq. 5. The greater current steering angle and the lower the driving speed, the higher the score for this point.

$$imp_i = ste_i \circ \left(\frac{1}{spd_i + 1}\right) \circ dis_i \qquad (5)$$

where the $\circ$ indicates the concatenation operator.

The attention weight gotten from the scorer layers can be further calculated according to the Eq. 6.

$$a_i = ReLU(W_a \cdot imp_i + b_a)$$
$$u_i = \exp(a_i)/\sum_1^t \exp(a_i) \qquad (6)$$
$$Attn_{cur} = \sum_1^t u_i \cdot h_i$$

where $ReLU$ is the activation function, W_a, b_a represent the weight and deviation of the mapping matrix respectively, and u_i is the normalized weight vector.

Location Prediction Sub-module. The sub-module outputs a geographic coordinate with specific latitude and longitude as predicted destination. It is achieved by the residual blocks in our model. The residual blocks use the historical sequence score and the current trajectory score as the initial input, pass through the activation function and the fully connected layer residual function in turn, and finally output a vector of size two according to Eq. 7.

$$ST = Attn_{hist} \circ Attn_{cur}$$
$$x_0 = \sigma_r(ST)$$
$$x_i = \sigma_r(\sigma_r(x_{i-1}) + x_{i-1}) \quad i > 0 \qquad (7)$$
$$result = f(\tanh(x_4))$$

where $Attn_{hist}$ represents the attention score of the historical sequence, $Attn_{cur}$ represents the attention score of the current trajectory, the ∘ indicates the concatenation operator. σ_r represents the fully connected residual function. The f function represents a linear function that converts 128-layer input to 2-layer output.

4 Experiments

In order to verify the effectiveness of our model and its modules, two comparative experiments are conducted. Meanwhile, we also discuss performance influences of importance factors on prediction including grid size, historical sequence time span and travel distance.

4.1 Experimental Setting

Experiment Data. The experimental data used in this paper is from the trajectories of private car drivers in Shenzhen collected by built-in GPS devices for the year of 2018. This paper uses travel entropy to quantitatively evaluate the randomness and predictability of driving behavior. According to travel entropy, travel distance and number of trajectories, we selected four groups of data as sample data to verify the robustness of our model. The statistics of the data and visualized trajectories are shown in Table 1 and Fig. 4 respectively.

Table 1. Yearly travel statistics of the four drivers.

Driver ID	Yearly travel OD entropy	Number of different origins/destinations	Total travel distance (km)	Number of trajectories	Average travel distance per trajectory (km)
Driver_1	6.7	263	11554	1940	5.96
Driver_2	10.3	975	38717	2515	15.39
Driver_3	6.0	213	38588	3647	10.58
Driver_4	10.3	1272	47951	2618	18.32

Table 1 and Fig. 4 show that driver 1 has similar entropies with driver 3 but lower travel distance that we can analyze the impact of travel distance on prediction. In addition, driver 2 and driver 4 have higher entropies compared with driver 1 and driver 3, which can discuss the impact of travel uncertainty on prediction.

Experimental Environment and Parameter Settings. The model and comparative experiments are implemented on the Windows 10 using PyTorch 1.3.1, equipped with a NVIDIA GeForce GTX 960M GPU and 4 CPU (i5-6300HQ) cores. The model training process runs on the CUDA device, and the data is divided into the training set and the test set at a ratio of 4 to 1. The batch size we set is 64 and the epochs are 50. The trajectory

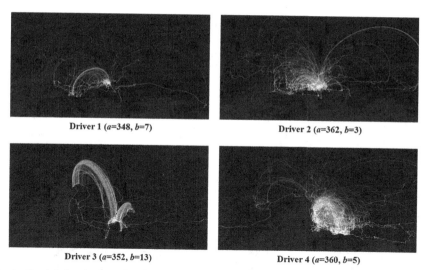

Fig. 4. Spatial distribution of trajectories. Where *a* in caption of subfigures refers to the number of historical OD sequences and *b* refers to the number of missing sequences. The starting point and the destination of the driving trip are connected by arcs, indicating a segment of the driver's trip.

points are gridded, the grid is mapped to a vector which size is R^{20}, the time is mapped to a vector which size is R^{20}, day_{sem} and $week_{sem}$ are mapped to R^{10}, R^5. The dimension of hidden state in LSTM is 128, and dropout layers' parameter is set to 0.5 to prevent model overfitting. Adam optimization algorithm is adopted to train our model.

Evaluation Metrics. Three criterions are used to evaluate model accuracy, including mean absolute error (MAE), root mean squared error (RMSE), and mean relative error (MRE). MAE evaluates the predictive ability of the model, RMSE reflects the stability of the model's prediction accuracy. MRE evaluates the relative prediction accuracy of different distances. Each evaluation metrics is shown in Eq. 9.

$$
\begin{aligned}
MAE &= 1/n \sum_{(i=1)}^{n} |Dis(y_i, \hat{y}_i)| \\
MSE &= (1/n \sum_{(i=1)}^{n} (Dis(y_i, \hat{y}_i))^2 \\
MRE &= 1/n \sum_{(i=1)}^{n} \frac{|Dis(y_i, \hat{y}_i)|}{total_{dis_i}}
\end{aligned}
\tag{9}
$$

where n is the number of trajectory sequences, $Dis(y_i, \hat{y}_i)$ is the distance difference between the real destination y_i and the predicted destination $\hat{y}_i$, $total_{dis_i}$ represents the total travel mileage of the track sequences.

4.2 Performance Comparison

Validation of Robustness and Validity of Model. The study selected three baseline models, i.e., hidden Markov model (HMM), Random Forests (RF), and basic LSTM. The LSTM is commonly used to deal with long sequence problems. The HMM model can better obtain the state transition probability between targets which has a good performance in the processing of sequence and probability problems. The RF has a simple structure, high efficiency and accuracy for multi-dimensional data processing and can effectively solve the problem of overfitting. The experiment results are shown in Fig. 5 and Table 2.

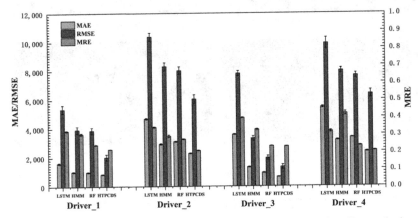

Fig. 5. The accuracy comparison of the proposed model and three baseline methods.

Table 2. The performance of the four models on the four selected drivers.

Driver ID	Methods	Average and standard deviation (SD (*))					
		MAE/(m)	SD(MAE)	RMSE/(m)	SD (RMSE)	MRE	SD (MRE)
Driver_1	LSTM	1614.598	43.286	5370.166	276.386	0.321	0.003
	HMM	1012.422	31.881	3932.357	231.763	0.303	0.005
	RF	989.693	32.001	3872.093	219.368	0.24	0.002
	HTPCDS	834.774	26.724	2022.236	200.149	0.214	0.001
Driver_2	LSTM	4695.326	45.867	10381.25	272.083	0.343	0.003
	HMM	2930.188	41.309	8304.159	273.717	0.29	0.007
	RF	3103.72	31.356	8022.203	261.96	0.273	0.002
	HTPCDS	2295.295	27.58	6049.745	292.338	0.207	0.002
Driver_3	LSTM	3600.26	33.695	7817.886	173.508	0.397	0.002
	HMM	1322.081	27.917	3313.72	168.92	0.329	0.003
	RF	931.43	26.271	1960.342	150.757	0.232	0.001
	HTPCDS	646.3	15.104	1340.936	153.506	0.23	0.001
Driver_4	LSTM	5482.625	60.094	9908.601	393.031	0.317	0.006
	HMM	3200.248	36.227	8011.196	183.613	0.42	0.011
	RF	3381.004	18.037	7663.21	171.112	0.233	0.002
	HTPCDS	2392.335	29.488	6384.938	268.764	0.205	0.002

In general, our model has better performance compared with the three baselines. The traditional LSTM model performs worst because of the features which are hard to learn without attention mechanism. RF and HMM perform better than the traditional LSTM but are limited to the complexity of input data, because they can hardly capture travel features in long term sequences and mine the hierarchical temporal driving patterns. More specifically, our model has high accuracy and stability for Driver 1 and Driver 3 which have low entropies, and obtains higher MAE and RMSE in Driver 2 and Driver 4 that have high entropies. It proves that the travel uncertainty can affect the prediction performance. Meanwhile, although Driver 1 and Driver 3 have close travel entropies, their MAE, RMSE, and MRE under HTPCDS are significantly different. In general, Driver 3 has lower MAE and RMSE but slightly higher MRE than Driver 1. According to Table 1, we can find that Driver 3 has longer average travel distances but more samples. It indicates that larger sample size has more positive impact on the error reduction, and long travel distance does not necessarily associate with larger uncertainty and unpredictability.

Validation of Hierarchical Attention Mechanism and Current Traveling Context.
To verify the influence of historical travel patterns and stacked frames of attention mechanism on prediction accuracy, we conducted comparative experiment upon historical OD sequences (LSTM+OD), single time attention (TA), hierarchical temporal attention (HTA). We furtherly compare HTPCDS with HTA to verify the effectiveness of current traveling context. The experiment results are shown in Fig. 6 and Table 3.

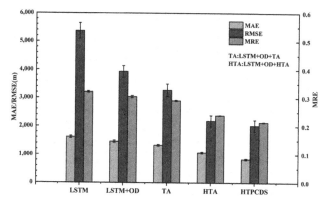

Fig. 6. The accuracy comparison of hierarchical attention mechanism and current traveling context.

In general, experimental results show that the historical sequences and the attention mechanism can effectively improve the prediction accuracy and stability. Compared with the traditional LSTM, the accuracy with historical OD sequences is improved slightly because the model is not capable enough to learn the historical patterns in OD sequences. In contrast, it can be improved by utilizing the attention mechanism. The attention mechanism of the double-layer structure is better than the single-layer attention mechanism in terms of the accuracy and stability. Because double-layer structure learns calendar period on both daily and weekly basis and can restore the state of motion in a

Table 3. Comparison of prediction accuracy results of experimental models.

Methods	Average and Standard Deviation (SD (*))					
	MAE/(m)	SD(MAE)	RMSE/(m)	SD (RMSE)	MRE	SD (MRE)
LSTM	1614.598	43.286	5370.166	276.386	0.321	0.003
LSTM+OD	1463.015	40.319	3937.53	203.613	0.304	0.004
TA	1329.909	29.131	3272.569	221.756	0.29	0.002
HTA	1063.733	27.627	2187.436	196.388	0.238	0.001
HTPCDS	834.774	26.724	2022.236	200.149	0.214	0.001

long term of time more effectively. Meanwhile, we found that HTPCDS had lower error than HTA because the spatial correlation in current driving status are mined by spatial learning sub-module.

4.3 Impact Factor Analysis

Grid Partition Granularity. The comparative experiment with increasing grid sizes is to discuss the impact of input sequences under different grid division granularities. The experiment results are shown in Fig. 7.

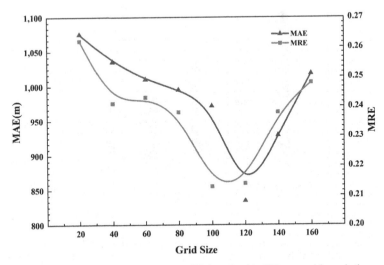

Fig. 7. Average prediction accuracy of Driver 1 with different grid resolutions.

Figure7 shows that grid granularity affects the prediction performance. The coarse grid division, such as 20 by 20, can easily cause the trajectory points in different areas to be allocated to the same grid ID, which may classify irrelevant OD into the same category. Similarly, the fine-grained grid division, such as 160 by 160, restricts the ability to reflect the spatial correlation and performs poorly in accuracy. Therefore, to obtain better prediction accuracy, dedicated experiment is desired to select appropriate gird granularity. The best grid size obtained in this paper is around 120 by120.

Time Span of Historical Sequence. This experiment is to verify the influence of the time span of historical sequence on the prediction performance. The changing trend of MAE under different time spans of historical OD sequences are shown in Fig. 8.

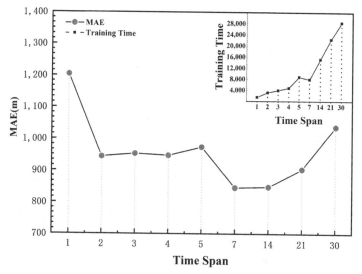

Fig. 8. The prediction accuracy and training time of historical sequences in different time spans

The experiment results show that the time span chose one week is the best which is consistent with weekly-based travel activities. The time span of two weeks also has similar performance on prediction results. However, the time span more than two weeks performs worse which may be related to our historical temporal attention module structure. The historical module is stacked by two attention layers that are used to learn daily and weekly relationships. It has limited capacity to mine monthly patterns. Meanwhile, the time span of a day has a high error because only day attention layer in historical temporal attention module can be used. Furthermore, the time span affects the training time of the model. The larger the time span of model selection, the longer the time required for model training.

Travel Distance. The prediction performance is affected by the current travel distance within a single trip. We conducted comparison experiments under different distances to verify the impact of current travel distance.

As shown in Fig. 9, MAE presents an upward trend but MRE shows a downward trend in general. We found that the longer travel distance and smaller travel frequency can lead to the uncertainty of model prediction. More specifically, the sub figure shows that

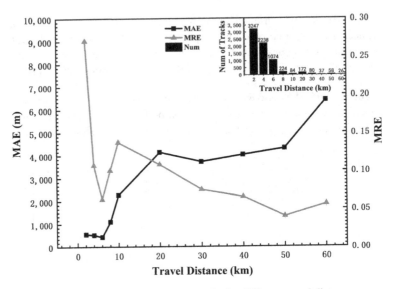

Fig. 9. Experimental evaluation results for different travel distances

the short-distance travel behaviors are frequent, and the number of travel trajectories smaller than 6km occupies a large proportion of all travel trajectories, which MAE performs well but MRE performs much higher due to the shorter travel distance. When the travel distance is between 8km and 10km, the MAE of the model prediction increases rapidly due to the small number of travel trajectories, and the MRE shows a trend of rebound because the current travel distance is still relatively short. Therefore, travel distance and the number of trajectories affect the prediction performance of the model. In order to solve the problem of sparse long-distance travel data, it is possible to enrich the data of each travel distance by increasing the number of sub-trajectories during data preprocessing.

5   Conclusion

This paper proposed a real-time driving destination prediction model, HTPCDS, based on historical travel patterns and current driving status. HTPCDS is constituted by the two modules, i.e., the historical temporal attention module and the status-aware prediction module. The historical attention module stacked by two attention layers extracts historical patterns in calendar period. More specifically, daily attention layer pays attention to the correlation between track points in a day, and weekly attention layer focuses on the importance of the days in a week. While, the status-aware prediction module mines the spatial pattern according to the current travel context. Through comparative experiments on the trajectory data of four drivers, the effectiveness and robustness are verified. Meanwhile, we further discuss the impact factors which affect our model performance including current travel context, historical time span, travel frequency and distance.

The model can be improved from the following perspective. Firstly, our model only considers the location and timestamp information of the OD point. Further research will integrate semantic-level feature (such as Urban functionalities of POI) and road network structure to solve the feature sparse problem of historical OD sequences. Moreover, our paper adopts a method by gridding geographic planes to process geographic coordinates, which may affect prediction accuracy. We will adopt the spatial clustering to replace the grid, which may enhance the ability to express the historical activity chain of a driver. Meanwhile, it will benefit model migration by turning the transfer of adjacent geographic coordinates into a span of adjacent functional intervals.

References

1. Zhang, H., Li, Z.: Weighted ego network for forming hierarchical structure of road networks. Int. J. Geograph. Inf. Sci. **2**(25), 255–272 (2011). https://doi.org/10.1080/136588109 03313534
2. Huang, Q.: Mining online footprints to predict user's next location. Int. J. Geograph. Inf. Sci. **3**(31), 523–541 (2017). https://doi.org/10.1080/13658816.2016.1209506
3. Yang, M., Zhang, M., et al.: Neural machine translation with target-attention model. Inst. Electron. Inf. Commun. Eng. **3**(E103.D), 684–694 (2020). https://doi.org/10.1587/transinf. 2019EDP7157
4. Gao, Y., Jiang, D., Xu, Y.: Optimize taxi driving strategies based on reinforcement learning. Int. J. Geograph. Inf. Sci. **8**(32), 1677–1696 (2018). https://doi.org/10.1080/13658816.2018. 1458984
5. Xia, L., Huang, Q., Wu, D.: Decision Tree-Based Contextual Location Prediction from Mobile Device Logs. Mobile Inf. Syst. 1–11 (2018). https://doi.org/10.1155/2018/1852861
6. Li, F., Gui, Z., et al.: A hierarchical temporal attention-based LSTM encoder-decoder model for individual mobility prediction. Neurocomputing **403**, 153–166 (2020). https://doi.org/10. 1016/J.NEUCOM.2020.03.080
7. Bahdanau, D., Cho, K., Bengio, Y.: Neural machine translation by jointly learning to align and translate. In: International Conference on Learning Representations, ICLR 2015 (2015)
8. Altaf, B., Yu, L., Zhang, X.: Spatio-temporal attention based recurrent neural network for next location prediction. In: 2018 IEEE International Conference on Big Data, Big Data, pp. 937–942(2018). https://doi.org/10.1109/BIGDATA.2018.8622218
9. Xue, H., et al.: A location-velocity-temporal attention LSTM model for pedestrian trajectory prediction. IEEE Access **8**, 44576–44589 (2020)
10. Luong, M., et al.: Effective approaches to attention-based neural machine translation. In: Proceedings of the 2015 Conference on Empirical Methods in Natural Language Processing, pp. 1412–1421 (2015)
11. Vaswani, A., Shazeer, N., et al.: Attention is all you need. In: Neural Information Processing Systems 2017, NIPS, pp. 5998–6008 (2017)
12. Gui, Z., Sun, Y., et al.: LSI-LSTM: An attention-aware LSTM for real-time driving destination prediction by considering location semantics and location importance of trajectory points. Neurocomputing, (2021). https://doi.org/10.1016/j.neucom.2021.01.067

Dynamic Shortest Path Queries over Moving Objects on Road Networks

Siyu Chen and Jianqiu Xu[⊠]

Nanjing University of Aeronautics and Astronautics, Nanjing, China
{siyu,jianqiu}@nuaa.edu.cn

Abstract. With the development of intelligent mobile devices and communi-cation technology, moving objects have become a hot research topic. This paper studies the moving objects on the road network, and a dynamic shortest path query algorithm is proposed to solve the shortest path calculation between two moving objects on the road network. The algorithm is divided into pre-processing phase and query phase. The pre-processing phase pre-calculates and stores a small part of the shortest path according to the important positions of the road network and the pre-storage coefficient k. Then the pre-processing information is used to query the shortest path during the query process. Experiments are performed by using the road network data (nearly 200,000 segments) of Nanjing, including the setting of the pre-stored coefficient k, the pre-stored path size, and the road network size. The results show that the MOSP algorithm proposed in this paper outperforms baseline approaches by 2–3 orders of magnitude.

Keywords: Dynamic shortest path query · Road network · Moving object

1 Introduction

Due to the rapid development of mobile communication technology and positioning technology, querying moving objects on road network has become one of the most popular topics in the field of spatio-temporal data retrieval.

With the increasing availability of accurate geo-position, e.g., GPS, it is much more convenient to collect large populations of moving objects. In these situations, it is essentially important to study the shortest path problem between two moving objects. Some researchers consider shortest path queries over dynamic sub-graphs [1] or dynamic net-works [2]. So far little effort has been devoted to study the shortest path query between two moving objects on road networks, which is a challenging problem. Due to the expansion of the scale of urban road network and the demand for real-time planning of personal journey, some classical shortest path query algorithms cannot be used to solve the problem caused by large-scale data sets and short response time, such as Dijkstra and A* algorithm [3]. Therefore we need a more efficient and real-time algorithm for dynamic shortest path query between moving objects on road network.

Dynamic shortest path query is widely used in our life. For example, in Didi, the position of passengers and drivers changes in real time. We should dynamically plan

© Springer Nature Switzerland AG 2021
X. Meng et al. (Eds.): SpatialDI 2020, LNCS 12567, pp. 44–59, 2021.
https://doi.org/10.1007/978-3-030-69873-7_4

a shortest path to ensure that passengers can get in the car as soon as possible. In addition, the query can be widely used in many fields such as traffic navigation, real-time navigation, automatic driving, etc.

In this paper, we study the problem of dynamic shortest path query for moving objects on road network. We use an example to describe this problem (see Fig. 1). Suppose that Car1 and Car2 drive along their respective road sections at the speed v_1 and v_2 respectively. When they reach the next road section, the speed of the two vehicles becomes v_1', v_2'. In this example, we default that the car is driving at a constant speed in the same road section, and the speed will change only when they switch the road section. The query needs to return the shortest path between two cars in real time. The basic method is to calculate the location of two cars at the same time per 1s and repeat the shortest path calculation. Assuming that the two cars are far apart, the calculation is very large. However, most of the calculations are repeated. As long as the car does not change the road section, the shortest path will not change much, which transform a dynamic problem into a static problem. The shortest path only needs to be recalculated when the moving object switches the road section. The small part of the change on the road section can be obtained through a time-distance function.

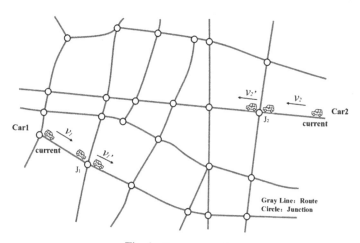

Fig. 1. Query example

Our main contributions in this paper can be summarized as follows:

1) We study a new problem of the dynamic shortest path query for moving objects on road network, which is significant to many real-world location-based applications;
2) We propose CMOSP algorithm base on pre-storage to reduce repeated calculation. We introduces the split point to ensure the accuracy of the query results, and a new storage structure to reduce the cost of pre-storage.
3) We evaluate the effectiveness and efficiency of our algorithm with the data set of Nanjing road network (nearly 200000 segments). We mainly adjust the pre-storage coefficient k, and analyze the influence of pre-storage structure, road network scale, distance between moving objects on the query performance. The results show that

the proposed CMOSP algorithm improves the query efficiency by 2–3 orders of magnitude compared with the basic method.

The rest of the paper is organized as follows. The related work is discussed in Sect. 2. Section 3 defines the problem. Our CMOSP algorithm is introduced in Sect. 4. In Sect. 5, the evaluation is reported. We conclude the paper in Sect. 6.

2 Related Work

The shortest path query in road network environment can be divided into the following three kinds of queries according to the attributes of query objects, query between (i) two static objects; (ii) one static object and one moving object; (iii) two moving objects.

The third query is the most difficult but the most common one. The first two queries are special cases of the third query. The difficulty of the query lies in two aspects. On the one hand, the large-scale road network increases the complexity of the shortest path calculation. On the other hand, when the location of the two objects changes, the shortest path also changes, and the calculation times increases. At the same time, the shortest path needs to be updated dynamically and efficiently. Therefore, how to reduce the repeated calculation is the key of the research.

At present, there are several solutions to reduce the repeated calculation, such as approximate calculation method [4], landmark method [11–13], pruning method [10] and hierarchical road network division method [5–9]. Among them, approximate calculation needs to sacrifice accuracy for efficiency. We mainly discuss the last three methods because this paper discusses accurate calculation. The idea of hierarchical road network division is to use the shortest path between sub networks to reduce the cost of path calculation in the process of query. HEPV [5] divides the graph by cutting the vertices, and computes all the shortest paths between all boundary pairs in advance. Storing all such pairs is time-consuming and takes up space, and cannot support large graphs. ROAD [6] recursively divides the whole graph into hierarchies of interconnected regional subnets. The shortest paths between any two points of the road network can be obtained quickly by using the shortcuts. G-tree [7] is a hierarchical structure for KNN query. G-tree also supports single source shortest path query. A component-based method is proposed to effectively calculate the shortest path between two vertices. [8] proposes an index based on hierarchical graph partition, which is called partition tree. Delling [9] provides a distance query technology, which uses the hierarchical hub labeling technology to determine the distance between two vertices on the network. The technology is robust to the network structure, can be extended to large networks.

HiTi [10] is the application of pruning method. This method uses A* algorithm to calculate the shortest path of objects in different subgraphs. The method uses Euclidean distance to estimate the lower limit of road network distance, and then uses A* algorithm to prune subgraphs with large distance.

The basic idea of landmark method is to find the important points on the road network, and speed up the search through the corresponding pre-calculation between the important points. TNR [11] calculates the distance from each vertex to a group of transport nodes, and uses transport nodes to calculate the shortest path distance. CH [12] computes the road network in advance by adding additional edges. Then CH uses bi-directional shortest path search to limit the edge to induce to more important nodes, which reduces the search space. PHL [13] uses highway labeling and pre-processing algorithms to calculate distances. This provides an idea of pre-computing and storing parts of shortest paths for speeding up the query.

All of the above methods solve the shortest path query between static points, but they cannot be directly applied to the dynamic shortest path query between moving objects on the road network.

With regard to solve the problem of network moving object query, researchers also propose some effective methods to improve the query efficiency. [14] transforms the shortest path calculation problem of two moving objects on road network into the calculation of two road segments. [15] uses the approximation method to solve the continuous KNN query problem. The method divides the road network into hierarchical trees, and forms a multi-layer shortcut road network. When the initial point to be calculated is far away from the target point, the distance can be calculated through the hierarchical tree and the multi-layer shortcut network, so as to reduce the calculation cost. [16] proposes the V-tree structure, which is an optimized structure based on G-tree. V-tree uses the road network iterative division method and puts forward the concept of active vertex to associate the moving object with the nearest vertex in V-tree. When the location of the object is updated, only the tree nodes on the path need to be updated from the corresponding leaf nodes to the root. This index solves the KNN query of the moving objects.

In addition, the safe region method is also commonly used to solve queries in the continuous moving state. This method effectively avoids repeated queries by using the feature that the query results of moving objects in the safe region remain unchanged. When querying moving objects, it only needs to judge whether they are in the safe region. [17] uses the safe region method to narrow the query scope. In [18], a fast calculation method of safe exit point is proposed. However, both of the two methods are focus on a static point and a moving point, and the safe region is also generated according to the static object. When both of them are moving, the safe region will also change dynamically. This method is mainly used to solve the continuous KNN query. However, this method provides inspiration for the concept and calculation of the split point in this paper.

Through the analysis, we found that the above methods can not directly apply to the dynamic shortest path query between moving objects in large-scale road network. Firstly, the cost of shortest path calculation is very high due to the large scale of the road network. Secondly, for real-time query, if the location of moving objects is updated frequently, we need to constantly recalculate a shortest path. In this case, the real-time demand is hard to meet. Therefore, it is important to reduce the repeated calculation and speed up the query efficiency through proper pre-calculation.

3   Problem Formulation

In this section, we explain and define the problem of dynamic shortest path queries over moving objects on road networks.

Definition 1 (Road Network). We model a road network as a weighted undirected graph $G = <V, E>$, where V is a set of vertices, representing the junctions of the roads; E is a set of edges, representing the road sections. V is denoted as $V = \{(s_1, d_1, s_2, d_2, cc)|s_1 \neq s_2, d_1, d_2 \in double, cc \in int\}$, where s_1, d_1, s_2, d_2 respectively represent the identification of which two sections intersect, and where they intersect each other. cc is the sign of whether s_1 and s_2 are connected. E is denoted as $E = \{(s, len, curve)|s \in int, len \in double, curve \in line\}$, len represents the length of the segment. The relationship between the road sections and the junctions is shown in Fig. 2.

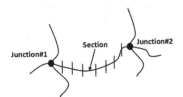

Fig. 2. Road section

Definition 2 (Moving Objects). Each moving object (e.g., vehicle) on the road network is represented as $m = <t, pos>$, where pos represents the geo-location of m at time t. Note the locations of moving objects is updated per second. Let $MO = \{m_1, ..., m_n\}$ denote a set of moving objects.

Definition 3 (Dynamic Shortest Path). Given a road network G, a shortest path set is denoted as $L(s, d)$, which is composed of all the shortest paths from the road section s of moving object m_1 to the road section d of moving object m_2 at current time t. If $CurDist(l, \delta, t) \leq CurDist(l', \delta', t)(l \in L, l' \in L-l)$, the real-time shortest path from m_1 to m_2 at the current time, denoted as $CPath(m_1, m_2, t)$, can be computed by $l + \delta$. $CurDist(l, \delta, t)$ is computed by the length of path l at time t plus the offset of m_1, m_2 on respective road sections. t changes within a time interval, which starts from the current moment and end with the moment that any one of the moving objects leaves the road section.

For example, in Fig. 1, $L(s, d)$ corresponds to the set of four shortest paths generated within the endpoints of the road sections where two cars are located at current. As the cars move on the road section, l will switch between the four shortest paths. The interval range of t indicates the interval from the current moment to the moment when Car1 leaves J_1 or Car2 leaves J_2. The query returns the dynamic shortest path corresponding to the time in the interval.

Definition 4 (Dynamic Shortest Path Query). Given road network G, a set of moving objects $MO = \{m_1, ..., m_n\}$ and a query $q = <m_i, m_j> (m_i, m_j \in MO)$, the answer of q is the dynamic shortest path between m_i and m_j, denoted as $CPath$.

4 CMOSP

Data pre-processing is a typical method of exchanging space for time. During the pre-processing, a lot of storage space will be consumed to pre-calculate and store the shortest path distance between all nodes. This will lead to memory overflow and query efficiency reduction. Therefore, we tend to process important information in advance, which reduces the storage cost and speeds up the query. However, this method also needs to meet two points: (i)pre-processing should be fast enough and the amount of data to be saved after pre-processing should be small enough; (ii)we must ensure that the shortest path can be accelerated in the real-time calculation process. At present, a lot of research work is devoted to finding a good compromise between the two points. In this paper, the dynamic shortest path algorithm proposed is a compromise between pre-storage cost and query performance.

In this section, we propose the CMOSP algorithm to solve the problem of dynamic shortest path query over moving objects on road network. The algorithm is divided into two stages: pre-processing and query. The main purpose of the pre-processing stage is to pre-calculate the shortest path between some important points, and find the relationship between the number of pre-stored path and the coverage of the road network. At the same time, we find the relative optimal coefficient, so as to reduce the calculation of the shortest path and improve the query efficiency.

G-tree index has a significant effect on improving the efficiency of the shortest path query on the road network. The index divides the road network into subnets and computes the shortest path between subnets in advance. However, in terms of space, both leaf nodes and non-leaf nodes need to store distance matrix. The disk space occupied by index and the maintenance cost of index updates are high. The method proposed in this paper aims to achieve the relative balance between space and time by storing only a specific part of pre-calculated path. The difficulty lies in how many pre-calculated paths should be saved. We determine the coefficient by experiment. In the query stage, some information stored in the pre-processing stage is used to assist query and return results.

4.1 Storage Structure

In order to speed up the query processing, we build some auxiliary structures in the system. As shown in the right side of Fig. 3, we use 2-D R-tree to index the road network. The lower B-tree stores the identifications of road sections contained in the corresponding MBR and related information. In order to reduce the query cost, a certain number of pre-calculated paths, denoted as Pre_Path, need to be maintained, as shown in the left side of Fig. 3. In this section, we focus on selecting some important points, executing the shortest path query between them and maintaining a certain number of shortest paths.

We construct an array to store the *id* set of road section. The specific information of the road section can be obtained according to these *ids*, which can be associated with the 2-D R-tree and B-tree. Because there is a large overlap between these pre-stored paths, we find the longest continuous common sub-paths of several paths and store them as an array, recorded as *LCS*. In this way, we can reduce the storage cost.

For the purpose of quickly judge whether the query object falls on the pre-calculated path, we design an array structure to store some distant point pairs. Only the point pairs consisting of each segment *id* of the previous part of the public sequence and the segment *id* of the end point are saved, as shown in Fig. 3 (e.g., $<Sa_1, Se_n>$, $<Sa_n, Se_n>$). The corresponding path of each point pair can be restored by a three-part structure. When querying, we only need to judge whether the section *id* pair of the sub-trajectory pair can be found in the pre-stored point pair set. If yes, we can directly get the shortest path between the two sections. The final result is composed of this shortest path and the offset of the two objects moving on the starting and ending segments. Most of the repeated calculation is omitted.

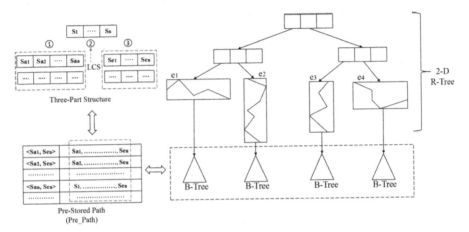

Fig. 3. Storage structure

4.2 Pre-processing

In the pre-processing stage, the main considerations are the selection of important points and the determination of the number of pre-stored paths. As for the important points, we want to select the vertices with high connectivity and far away from each other. However, the vertices with high connectivity are generally clustered, so that the pre-computed path will only cover a small part of the road network. If only a single indicator is considered, the important points in the road network cannot be accurately and effectively explored. Therefore, we should consider the global impact of nodes in the network. The selection of important points in this paper is determined by considering the topology of the road network. We mainly combine two factors: (i) the number of neighbor nodes; (ii) the distance from the center of the road network. The two factors are weighted separately, and the evaluation indexes I_i of node importance are computed as

$$I_i = \alpha \bullet \sum_{j=1, j \neq i}^{n} \delta_{ij} + (1 - \alpha) \bullet dist(v_i, v_{mid}) \qquad (1)$$

$\sum\limits_{j=1,j\neq i}^{n} \delta_{ij}$ represents the connectivity of each node, δ_{ij} represents whether node v_i and v_j are adjacent. If they are adjacent, $\delta_{ij} = 1$, otherwise, $\delta_{ij} = 0$. The formula refers to the degree centrality principle, which is the most direct factor to evaluate the importance of a node. $dist(v_i, v_{mid})$ represents the road network distance between node v_i and central node v_{mid} of road network. We give $\sum\limits_{j=1,j\neq i}^{n} \delta_{ij}$ a weight value α and give $dist(v_i, v_{mid})$ a weight value $1-\alpha$. We need to maintain as few pre-computed paths as possible, but cover as much space as possible. Combined with the experiments and previous experience, it is found that when the node connectivity occupies a larger weight in the evaluation index, the selected point has better effect. In this paper, we set α to 0.8. We select some important points of road network, and expand them to an intersection. Then we maintain the pre-calculated paths between these points, a total of n, and sort these paths according to the length from high to low. Only $O(k \cdot n)$ paths are retained, where k refers to the pre-storage coefficient.

4.3   Query Procedure

The query phase is performed in two steps:

1) Determine the *ids* of the road segments where the two moving objects fall on respectively, and judge whether the shortest path between the two road segments has been pre-calculated;

2) If the moving objects just fall on the pre-calculated paths, we can directly get a shortest path in the pre-stored information according to the road segment *id*. The shortest path at the current time is computed by summing up path above and the small segments from the current positions of the moving objects to the corresponding vertices of the road segments. If the moving objects do not fall on the pre-calculated paths, the shortest path between them needs to be recalculated. We can judge which endpoint of the road segment should the moving object leaves from by calculating the split points. The split points is defined in Definition 5.

In the interval from the current moment to the moment when the moving object leaves the road section, the complete shortest path between two moving objects does not need to be repeatedly calculated at all times.

Judge Whether the Moving Objects Fall on the Pre-calculated Paths:

1) The segment *id* sequence of the pre-stored path is divided into a three-part structure: the three-part structure is saved into a relational form with attributes (source, destination, path from source to *LCS*, *LCS*, path from *LCS* to destination). The *LCS* is saved in another structure.

2) Some distant point pairs are stored. Only the point pairs consisting of each segment *id* of the previous part of the public sequence and the segment *id* of the end point are saved, e.g., $< Sa_1, Se_n >$, $< Sa_n, Se_n >$.

3) When calculating the shortest path, we first determine whether the section *id* pair of the two moving objects can be found in the point pair set stored in 2). If the shortest path can be obtained directly, there is no need to repeat the calculation.

Definition 5 (Split Point). Given a pair of sub-trajectories after dynamic time warping, which fall on two road sections respectively. With objects moving, the shortest path will be arbitrarily switched among several shortest paths of the two road sections. The position where the change occurs is called split point, denoted as P_{split}.

The two paths between the road sections are fixed, and the only change is the little part of the object moving on the road network. When two objects do not reach the split point or junctions, the path does not change obviously. Assuming that the speed of moving objects on the same road segment is constant, these split points can be found by determining the distance function. The function of split point is to ensure the accuracy of calculation. We use split points to judge from which endpoint of the road segment should the moving object leaves during which time interval. Of course, the split point does not always exist. In other words, no matter how the moving objects move on some certain road sections, they should always leave from one endpoint. We calculate the split point to ensure that the path is always the shortest, but we do not need to recalculate the complete path at all times. We only need to judge whether the split point exists and whether the moving object moves to the front or the back of the split point.

Split Point Calculation:
As for the split point P_{split}, we explain the situation of a static object, a moving object and two moving objects respectively. For the convenience of description, only two cases of the shortest path between the endpoints of two sections are considered. For example, the endpoint A can only reach the endpoint D, and the endpoint B can only reach the endpoint C, as shown in Fig. 4.

Case 1. Query the dynamic shortest path between moving object O_1 and static object O_2. As Fig. 4 shows, we find that O_1 moves on the 300 m long road section AB, O_2 moves on the 180 m long road section CD. O_2 is 80 m away from D. Due to O_2 is static, when calculating the shortest path from O_1 to O_2, it involves whether the moving object O_1 is better to exit from A or exit from B. A split point can be found on road section AB. So that when the moving object O_1 is in front of the split point, we always choose exit A. On the contrary, we always choose exit B. This can ensure the accuracy of the shortest path calculation. According to Fig. 4, the shortest path distance between B and O_2 (B $\rightarrow$ C $\rightarrow$ O_2) is 500 m in total, the shortest path distance from A to O_2 (A $\rightarrow$ D $\rightarrow$ O_2) is 380 m in total. Therefore, the split point P_{split} on the road segment AB should lie in 210 m from A.

Case 2. Query the dynamic shortest path between moving object O_1 and O_2. On the basis of Case 1, we assume that O_2 is also moving. We set the speed of O_1 as v_1, the direction as A to B, the speed of O_2 as v_2, the direction as D to C. We assume that the speed of moving objects on the same road section is constant. l_1, l_2 are the length of road section AB and CD respectively. In this case, the split point also exists. When O_1 is between A and P_{split}, O_1 is better to leave through A to O_2. When O_1 is between

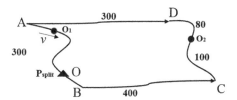

Fig. 4. Moving object and static object

P_{split} and B, O_1 is better to leave through B to O_2. As shown in Fig. 5, the shortest path distance between O_1 and O_2 is as follows:

a) $O_1 \rightarrow A \rightarrow D \rightarrow O_2$: $dis = v_1t + d_{origin} + d_1 + v_2t$;
b) $O_1 \rightarrow B \rightarrow C \rightarrow O_2$: $dis = l_1 - d_{origin} - v_1t + d_2 + l_2 - v_2t$.

d_{origin} represents the distance between the initial position P_0 of the moving object O_1 and the endpoint A. Therefore, the distance between P_{split} and A is computed as a function $d_{origin} + \frac{v1(l1+l2+d2-d1-2 \cdot d_{origin})}{2(v1+v2)}$, $d_{origin} \in [0, l_1]$.

After finding the split point, the complete shortest path between two moving objects does not need to be calculated repeatedly every second. We can determine the fixed path in the middle according to P_{split}, then plus a time-distance function.

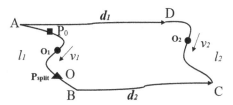

Fig. 5. Two moving objects

Algorithm 1 presents the details of our technique. Initially, we build 2-D R-tree and B-tree according to the road network, and initialize *CPath* and *L*. First of all, we determine the section e_1 and e_2 where the moving object m_1 and m_2 fall on (line 3). After obtaining the two section *ids,* we discuss in two situations by judging whether the *ids* exist in the point pair set of *Pre_Path* (lines 4–14). If there exist the *ids,* we can get a pre-computed shortest path between section e_1 and e_2 through *Pre_Path.Get(i)*. Then the final result *CPath* can be computed by *Pre_Path.Get(i)*, summing up with the offset segments of m_1 and m_2 on section e_1 and e_2. If the *id* pair cannot be found, we should use A* algorithm to recalculate several shortest path sets $L(e_1, e_2)$ for two sections. In this case, calculating P_{split} is an important step to reduce the computation cost. The calculation method *ComputeP$_{split}$()* has a detailed description at the definition of the split point, which is related to the moving object's moving direction and speed. After obtaining P_{split}, we determine whether the shortest path between the sections is l_1 or l_2 by judging the relative position of moving object m_1 and P_{split}. Finally, *CPath* equals to

l_1 or l_2 add up with the offset segments of m_1 and m_2 on section e1 and e2. The above is the whole process of the algorithm.

Algorithm 1 *CMOSP(m1, m2, Pre_Path)*

Input: m_1, m_2: moving objects; *Pre_Path*: pre_stored paths
Description:
1: build 2-D R-tree and B-tree according to road network
2: $CPath \leftarrow \emptyset, L \leftarrow \emptyset$
3: determine section e_1, e_2 where m_1, m_2 lie in
4: **if** (*Pre_Path.Find(<e_1.id, e_2.id>)*)
5:   $CPath(m_1, m_2, t) \leftarrow Pre\_Path.Get(i)$ + offset of m_1, m_2 on section e_1, e_2
6: **else**
7:   $L(e_1, e_2) \leftarrow A^*(e_1, e_2)$
8:   $ComputeP_{split}(m_1, m_2)$
9:   **if**(at time t, m_1 is before P_{split})
10:     $CPath(m_1, m_2, t) \leftarrow l_1$ + offset of m_1, m_2 on section e_1, e_2
11:   **else**
12:     $CPath(m_1, m_2, t) \leftarrow l_2$ + offset of m_1, m_2 on section e_1, e_2
13:   **end if**
14: **end if**
15: **return** *CPath*

---

5 Experimental Evaluation

5.1 Setups

The evaluation is conducted in a standard PC (Intel(R) Core (TM) i7-4770CPU, 3.40 GHz, 8 GB memory) running Ubuntu 14.04 (64 bits, kernel version 4.8.2–19). We develop the CMOSP algorithm in C/C+ + and integrate the implementation into an extensible database system SECONDO.

We use real road network datasets, as reported in in Table 1. SubNJ represents the subnet of Nanjing. SanF represents San Francisco.

Table 1. Road network dataset.

Road network	Sections	Junctions
Nanjing	182861	122013
SubNJ	3223	2162
Berlin	11440	8742
SanF	221415	175343

The moving object dataset adopts simulation data which randomly generated based on different road networks. The speed information of moving objects is generated by

Thomas Brinkhoff's moving objects generator. We set that moving objects move at a constant speed on the same road section, the life cycle of moving objects is 1 min, and the time interval of location update is 1 s. Four different sets of moving object datasets are generated for different road networks. Each group has 100 moving objects. The first two groups of moving objects are generated by randomly selecting points on the road network, and try to keep them from falling on the pre-stored paths, recorded as SET1 and SET2. The last two groups of moving objects are selected in a range, which is within 1km from the pre-stored paths, recorded as SET3 and SET4. So that the selected moving objects are more possible to fall on the pre-stored path.

This experiment is mainly divided into two parts for testing. First of all is the selection of the pre-storage coefficient k, and the second is the evaluation of the query performance of our algorithm.

5.2 Selection of Coefficient k

We set the pre-storage coefficient k to 1, 5, 10, 20, 30, 40, 50, 60. We take Nanjing road network as an example. Table 2 shows the change in the number of road segments covered by pre-calculated path as the coefficient k increases. We can analyze that in Nanjing road network dataset, when the coefficient is 10% and 20%, the number of road network segment coverage (CoverNum) increased year-on-year is relatively large. The purpose of storing as many road segments as possible with as less storage space can be achieved. Considering that the effect of 10% is not good enough, we prefer to set k to 20%.

Table 2. Impact of pre-storage factor on road network coverage.

k	60%	50%	40%	30%	20%	10%	5%	1%
CoverNum	4316	3888	3527	3162	3059	2084	1450	942

According to Fig. 6, when k is set to 20%, the memory consumption of using the three-part structure proposed in this paper will be 85% less than directly storing the entire path.

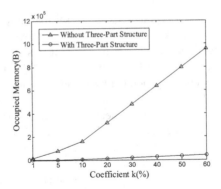

Fig. 6. Memory of pre-storage

5.3 Query Performance

At present, there is little research on dynamic shortest path query over two moving objects on road networks. At the same time, there is no special algorithm for comparison. Therefore, our baseline algorithm in this experiment adopt an extension of the traditional A* algorithm without pre-storage structure. The pre-storage coefficient k is set to 20% according to the experiment above. We compare our CMOSP algorithm with baseline method, and analyze the influence of three factors on the query response time. Three factors are the pre-storage structure, the scale of road network, and the distance between moving objects. In order to avoid the randomness of the experimental results, the average of 30 test results is selected for the query response time.

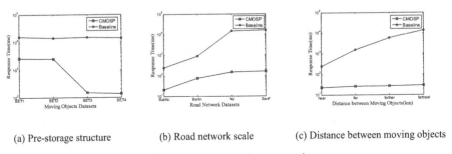

(a) Pre-storage structure   (b) Road network scale   (c) Distance between moving objects

Fig. 7. Query response time comparison

Pre-storage Structure. We take four groups of moving object dataset to test, which are SET1, SET2, SET3, SET4. The first two groups are generated by randomly selecting points on the road network, trying to keep them from falling on the pre-stored path. The last two groups are selected within the range of pre-stored path. By default, we take Nanjing road network and set k to 20%. Each query takes any two moving objects from each set for query.

Figure 7(a) shows the comparison of the two algorithms on query response time when using different moving object datasets. We can see that under SET1 and SET2 data sets, the decrease in the response time of the CMOSP algorithm compared with the baseline method is smaller than that of SET3 and SET4. Because the first two groups of moving objects do not fall on the pre-storage path, so we cannot use the pre-storage structure and need recalculate. However, it is better than baseline method. Because each time the position of the moving object is updated, in baseline method, the shortest path at the current position must be recalculated. Nevertheless, in CMOSP algorithm, only the little change on the road segment needs to be recalculated. We find that the average query response time of CMOSP algorithm in the latter two datasets is two orders of magnitude faster than that of baseline method. Thus we find that the pre-storage structure can significantly accelerate the dynamic shortest path query of road network moving objects.

Road Network Scale. We perform the evaluation by changing the road network scale. We take SET3 which is generated in the range of the pre-stored path as the moving object dataset, and set k to 20%.

We uses four different scale of road network datasets to test the query performance changes of the two algorithms. As shown in Fig. 7(b), we can see that the query response time of our algorithm varies little in different road network scale. While the response time of the baseline method will increase significantly with the increase of the road network scale. Under the small road network such as Nanjing subnet and Berlin, the query efficiency of our CMOSP algorithm has a certain improvement, but it is not obvious enough. Thus, the CMOSP algorithm proposed in this paper is more effective than the baseline method in large scale of road networks.

The Distance Between Moving Objects. In this part, we evaluate the effect of distance between moving objects on query performance. The pre-storage coefficient k is set to 20%, and the Nanjing road network is adopted. We take SET3 as the moving object dataset. In order to generate moving object pairs with different distances, the starting points of moving objects in SET3 are paired and sorted by distance. According to the sorted distance above, these objects are divided into four groups of query datasets with the same size, denoted as the near, far, farther and farthest moving object pairs.

According to Fig. 7(c), we observe that when the distance between moving objects changes, the performance of the CMOSP algorithm remains basically stable. However, as the distance increases, the baseline method needs to traverse more nodes to return the query results. In the CMOSP algorithm, as long as the mobile object pair can fall on the pre-calculated path, the query results can be obtained in a very short time, regardless of the distance.

6   Conclusions

In this paper, we present CMOSP algorithm to solve the dynamic shortest path query over moving objects on road networks. This algorithm proposes pre-storing a part of the shortest path to reduce the computational cost, and the split points are introduced to ensure the accuracy of the query results. The real road network data is adopted to carry on the experimental test, which mainly adjusts the pre-storage coefficient k, and makes an analysis of the influence of pre-storage structure, road network scale and distance between moving objects on query performance. The results show that the CMOSP algorithm proposed in this paper outperforms baseline approaches by 2–3 orders of magnitude.

Acknowledgment. This work is supported by NSFC under grants 61972198, Natural Science Foundation of Jiangsu Province of China under grants BK20191273 and National Key Research and Development Plan of China (2018YFB1003902).

References

1. Hassan, M.S., Aref, W.G, Aly, A.M.: Graph indexing for shortest path finding over dynamic sub-graphs. In: SIGMOD, pp. 1183–1197 (2016)
2. Huang, B., Wu, Q., Zhan, F.B.: A shortest path algorithm with novel heuristics for dynamic transportation networks. Int. J. Geogr. Inf. Syst. **21**(6), 625–644 (2007)
3. Goto, T., Kosaka, T., Noborio, H.: On the heuristics of A* or A algorithm in ITS and robot path-planning. In: IEEE/RSJ International Conference on Intelligent Robots & Systems. IEEE (2003)
4. Zhang, Z., Liu, J., Qiu, A., et al.: The shortest path approximation algorithm for large scale road network. Acta Geodaetica et Cartographica Sinica **48**(1), 86–94 (2019)
5. Jing, N., Huang, Y.W., Rundensteiner, E.A.: Hierarchical optimization of optimal path finding for transportation applications. In: CIKM, vol. 96, pp. 12–16 (1996)
6. Lee, K.C.K, Lee, W.C., Zheng, B.: Fast object search on road networks. In: Proceedings of the 12th International Conference on Extending Database Technology: Advances in Database Technology, pp. 1018–1029. ACM (2009)
7. Zhong, R., Li, G., Tan, K.L., et al.: G-Tree: an efficient and scalable index for spatial search on road networks. IEEE Trans. Knowl. Data Eng. **27**(8), 2175–2189 (2015)
8. Wang, J., Kai, Z., Jeung, H., et al.: Cost-efficient spatial network partitioning for distance-based query processing. In: MDM, vol. 1, pp. 13–22 (2014)
9. Delling, D., Goldberg, A.V., Pajor, T., Werneck, R.F.: Robust distance queries on massive networks. In: Schulz, A.S., Wagner, D. (eds.) Algorithms - ESA 2014. Lecture Notes in Computer Science, vol. 8737, pp. 321–333. Springer, Heidelberg (2014). https://doi.org/10.1007/978-3-662-44777-2_27
10. Jung, S., Pramanik, S.: An efficient path computation model for hierarchically structured topographical road maps. IEEE Trans. Knowl. Data Eng. **14**(5), 1029–1046 (2002)
11. Bast, H., Funke, S., Matijevic, D., et al.: In transit to constant time shortest-path queries in road networks. In: Proceedings of the Meeting on Algorithm Engineering & Expermiments. Society for Industrial and Applied Mathematics, pp. 46–59 (2007)
12. Geisberger, R., Sanders, P., Schultes, D., Delling D.: Contraction hierarchies: faster and simpler hierarchical routing in road networks. In: McGeoch, C.C. (eds.) Experimental Algorithms. Lecture Notes in Computer Science. WEA 2008, vol. 5038, pp. 319–333. Springer, Heidelberg (2008). https://doi.org/10.1007/978-3-540-68552-4_24
13. Akiba, T., Iwata, Y., Kawarabayashi, K., et al.: Fast shortest-path distance queries on road networks by pruned highway labeling. In: 2014 Proceedings of the Sixteenth Workshop on Algorithm Engineering and Experiments (ALENEX). Society for Industrial and Applied Mathematics, pp. 147–154 (2014)
14. Yin, X., Ding, Z., Li, J.: Moving continuous k nearest neighbor queries in spatial network databases. World Congr. Comput. Sci. Inf. Eng. **4**, 535–541 (2009)

15. Zhu, C.J., Lam, K.Y., Cheng, R.C.K., et al.: On using broadcast index for efficient execution of shortest path continuous queries. Inf. Syst. **49**, 142–162 (2015)
16. Shen, B., Zhao, Y., Li, G., et al.: V-tree: efficient kNN search on moving objects with road-network constraints. In: IEEE 33rd International Conference on Data Engineering (ICDE), pp. 609–620. IEEE (2017)
17. Zeberga, K., Jin, R., Cho, H.J., et al.: A safe-region approach to a moving k-RNN queries in a directed road network. J. Circ. Syst. Comput. **26**(05), 1750071 (2017)
18. Cho, H.J., Kwon, S.J., Chung, T.S.: A safe exit algorithm for continuous nearest neighbor monitoring in road networks. Mob. Inf. Syst. **9**(1), 37–53 (2013)

Environment Classification for Global Navigation Satellite Systems Using Attention-Based Recurrent Neural Networks

Haichun Liu[1]([✉]), Minmin Zhang[1], Ling Pei[1,3], Wei Wang[2], Lanzhen Li[3], Changchun Pan[1], and Zeya Li[4]

[1] Shanghai Key Laboratory of Navigation and Location Based Services,
Shanghai Jiao Tong University, Shanghai 200240, China
`haichunliu@sjtu.edu.cn`
[2] Department of Weaponry-and-Control, Beijing 100072, China
[3] Shanghai Beidou Research Institute, Shanghai 201702, China
[4] Beijing Institute of Tracking and Telecommunications Technology, Beijing 10095, China

Abstract. In this paper, an environment classification method for Global Navigation Satellite System (GNSS) is presented. The goal of the study is to characterize the statistical properties of the historical GNSS data in certain typical environments, so that appropriate localization or navigation algorithms can be chosen to achieve better performances once any environments are recognized in real practice. We extract Dilute of Precision (DOP) value, Carrier-to-Noise Ratio (C/N) and Number of Satellite in View from NMEA-0183 data collected in three real typical environments to characterize the environments. Further, an attention-based Recurrent Neural Network (RNN) is constructed; the historical characteristics extracted above are fed into the RNN. Attention values are then calculated using real-time characteristics and the RNN output in each time steps. High dimensional features are then constructed by soft attention and are used as the input of a fully connected network for classification. The performance of proposed method on the classification task of three typical environments has significantly improvement compared to recurrent neural networks without attention mechanism, and achieves an average accuracy of 94% on the testing set.

Keywords: Attention mechanism · Recurrent neural networks · GNSS signal · Environment classification

1 Introduction

Global Navigation Satellite System (GNSS) is a collection of multiple navigation systems, including GPS, GLONASS, Galileo and Beidou system [1], whose main function is to provide all-weather high-precision real-time positioning and navigation services [2]. Currently, the application of GNSS has gradually expanded to economy, technology, military etc., almost all fields of social life [3]. The robustness of receiver directly affects the positioning accuracy and navigation performance of GNSS system. So how to improve the robustness of the receiver is an issue worthy attention.

© Springer Nature Switzerland AG 2021
X. Meng et al. (Eds.): SpatialDI 2020, LNCS 12567, pp. 60–71, 2021.
https://doi.org/10.1007/978-3-030-69873-7_5

GNSS signal is in Ultra High Frequency band, there are many factors that will lead to the degradation of GNSS signal receiver performance, such as multipath effect [4], occlusion, signal attenuation, blocking and so on. In order to improve the performance of signal receiver, many anti-interference algorithms have been proposed, but these algorithms are only for specific environments, and the real-time calculation of algorithms increase the power consumption of the receivers. If signal receiver can identify the current environment accurately and switch to appropriate algorithm, it can save the system computing resources more effectively and improve the performance under the changing environments [5].

There are two main types of environment recognition methods, one is using vision sensors [6], the other is using the measurements of wireless signals. Although the environment recognition algorithm using vision sensor has been studied for long time, it needs additional hardware and energy consumption. This paper prefers to use the measurements of wireless signals for environment recognition, especially the measurement of GNSS signals.

GNSS receiver provides basic characteristics of GNSS signal in RINEX and NMEA-0183 protocols [7], such as signal strength, dilution of precision(DOP) value etc., which are primarily used to recognize the environment. Gao H et al. [8] use a hidden Markov model for the classification task of indoor and outdoor environments based on the characteristics of GNSS signals. While indoor GNSS signals are weaker than outdoor ones obviously. For vehicular navigation applications in the city, Hsu LT et al. [9] propose a novel dynamic Bayesian network to distinguish whether a vehicular is above or below a viaduct using the characteristics mentioned above. However, this algorithm needs to measure the height of the viaduct in advance. Lighari R U R et al. [10] introduced a data classification method for analyzing the aspects of SNR for GNSS in real conditions. The GNSS signal is affected by many factors, multi dimensional information should be considered. Evgenii Munin et al. [11] extract the important features and patterns from GPS L1 C/A signal data using deep convolutional neural networks (CNN) for multipath detection, the signal is mapped as a 2D input image. But he didn't take into account the temporal correlation of the signals, it can't be applied to real environment data. An SVM classifier is used by Yuze Wang et al. [12] for environment recognition, but in order to remove the influence of other factors, this method needs to collect some signal characteristics in the environment of open area as baseline.

In view of the above problems, this paper extends the method mentioned above and proposes an attention-based RNN for the classification task of three typical environments using NMEA-0183 data collected in real environment. We believe that this method does not need to measure additional reference values and some other prior knowledge and can be integrated into GNSS receiver to improve the performance of localization. It uses the historical time-series features and real-time features of GNSS signal to extract advanced high dimensional features for environment classification, and achieves high accuracy on the testing set.

The rest of the paper is organized as follow: In Sect. 2, we introduce the collection process and preprocessing of data. And the details of environment classification algorithm based on the attention mechanism are described in Sect. 3. In the next section, the results of a comparative experiment and visualization are shown to verify the validity of proposed method. Finally, we draw some conclusions in Sect. 5.

2 Data Collection and Preprocessing

2.1 Data Collection

The data set used in this paper was collected by a signal acquisition vehicle shown in Fig. 1. It is equipped with several GNSS receivers, a high-precision calibration system, a 64-line radar, a panoramic camera and a data collection system.

Fig. 1. Signal acquisition vehicle

The urban environment and geographical environment such as buildings, viaducts, airport, tunnel, mountain, plain, river, etc. are considered as the main criteria to divide environments. Based on these criterion, we divide all these data into more than 20 categories. In each class both the data of intermediate frequency signal and NMEA-0183 message with static and dynamic carrier are collected in different weather and time. NMEA-0183 data is chosen rather than intermediate frequency signal data for it's more suitable for real-time environment classification. Then three typical environments: open area, urban canyon, tree shade are selected as the classification task. These three typical environments respectively represents excellent signal quality, signal attenuation and multipath. The example photos of these three typical environments are shown in Fig. 2.

Open area refers to the environment that no trees or tall buildings around, which has excellent signal quality and almost all the signals can be received in the sky.

Tree shade refers to the environment that trees on both sides of the road cover most of sky over the signal acquisition vehicle. And to some extent, the strength of almost all the signals is attenuated in this environment.

Urban canyon refers to the environment that numerous tall buildings around the road and where the average height of buildings is usually larger than 50 m and the ratio of

Fig. 2. Example photos of three typical environments (a) is the environment of open area, (b) is the environment of tree shade, (c) is the environment of urban canyon

the height to road width is usually larger than 2. And some signals are blocked and the multipath interference is serious in this environment.

The detail of the data collected in the three typical environments mentioned above is illustrated in Table 1:

Table 1. An overview of the data collected in three typical environments

Environments	Open area	Tree shade	City canyon
NMEA-0183 files	11	12	29
Sample points of NMEA-0183 data	8519	7537	29275
Proportion of data collected in morning	31.3%	43.2%	31.4%
Proportion of data collected in afternoon	45.4%	26.6%	41.6%
Proportion of data collected at night	23.3%	30.2%	27%

2.2 Data Preprocessing

Data Preprocessing Flow. Because not all the characteristics in NMEA-0183 data are useful for environment classification task, and the data in searching satellites phase is significantly different from the data in normal working phase, we need to do some data preprocessing such as feature selection, noise reduction, normalization and data slicing. The data preprocessing flow is shown in Fig. 3.

Fig. 3. Data preprocessing flow

Feature Selection. NMEA-0183 data contains characteristics such as latitude and longitude coordinates, altitude, DOP values, satellite number, the elevation, azimuth and carrier-to-noise ratio of each satellite and so on. However not all these features are usable, some of which may even mislead classification model. Considering the correlation of environmental characteristics and generalization performance of the environment classification algorithm, DOP values, carrier-to-noise ratio of each satellite and the number of satellites in view are selected for representation of environment. However, the number of satellites in view is uncertain, if we allocate each satellite of GPS/GLONASS/Galileo/Beidou a channel, the feature matrix to represent environment will be sparse, which is hard for model training and generalization. So some statistics value of the carrier-to-noise ratio of satellites are extracted as features such as mean, standard deviation, median, upper quartile, lower quartile, maximum, minimum. Then together with the PDOP value, HDOP value, VDOP value and the number of satellites in view, it constructs the feature vector with 11 dimensions to represent environmental characteristics of each sample time.

Noise Reduction. The data in searching satellites phase is significantly different from the data in normal working phase, and the latter is more of our concern. So the data in searching satellites phase can be considered as noise, and we delete them for noise reduction. Sometimes the receiver may lose its position in normal working phase, which may lead the data during that period to be similar with the data in searching satellites phase. We can't just delete these data mentioned above, which may change the temporal characteristic of data. If we do nothing to these data, it will be harmful to the normalization, because the default value is much larger than normal value. So we change the default value of the data in normal working phase to a lower one, which is still larger than normal value.

Normalization. Normalization is an essential step of machine learning algorithm for the basic assumption of the input data is independent and identical distribution. Normalization can make sure features are on a similar scale which may accelerate the convergence and improve the performance of model. The normalization method this paper used is the z-score standardization, which normalize the original data of each feature satisfying the conditions to 0 mean and 1 variance. The normalized data z can be calculated as follows:

$$z = \frac{x - \mu}{\sigma} \tag{1}$$

Where x is the original data, μ is the mean of original data, σ is the standard deviation of original data. This will adjust the data of each dimension to the same scale, so as to avoid that the parameters trained by the model will focus on some kinds of dimension, and make the gradient descent converge faster.

Data Slicing. Usually the data collected in one acquisition process is over tens of minutes. And we don't need to feed all data into classification model at once which may lead to redundant calculation and even weaken the performance of classification model. So based on the assumption that each time period of the data with same label can effectively represent the environmental characteristics, this paper slices the original data into pieces with equal-length fragments to construct the data set. Specifically, the sliding window with sliding window width of 257 and step of 1 is used in this paper. Then each sample data in the data set is a $T \times C$ matrix, T means the number of sampling points, specific value is 257, where C means the characteristic channel, specific value is 11.

3   Environment Classification Algorithm

3.1   Attention Mechanism

RNN and its variants such as LSTM (Long Short-Term Memory) [13], GRU (Gated Recurrent Unit) [14] etc., are usually used to deal with time series data in deep learning methods. Sometimes the output feature vector of RNN may mismatch with the length of input data and it may forget some memory and contain some noise which make the output feature vector not the best representation of input data. Attention mechanism [15] is introduced to search a better representation in the field of NLP (Natural Language Processing), especially for the machine translation tasks with encoder-decoder structure.

Attention mechanism can be described as mapping a query and a set of key-value pairs to an output, where the query, keys, values, and output are all vectors. The output is computed as a weighted sum of the values, where the weight assigned to each value is computed by a compatibility function of the query with the corresponding key [16]. So the nature of attention mechanism can be regarded as using weighted summation to get a better representation for task. The key of attention mechanism is the compatibility function to compute the weight of each key-value pair. The general methods of calculate the similarity between query and keys are dot, general, concat, perceptron, which are computed as follow:

$$f(Q, K_i) = \begin{cases} Q^T K_i & \text{dot} \\ Q^T W_a K_i & \text{general} \\ W_a[Q; K_i] & \text{concat} \\ v_a^T \tanh(W_a Q + U_a K_i) & \text{perceptron} \end{cases} \tag{2}$$

Where Q means the vector of query and K_i is the i-th vector of keys, W_a, U_a and v_a^T are the parameter of attention layer.

Softmax function is usually used to calculate the weight of each value based on the similarity between query and keys. The specific formula is as follows:

$$a_i = \frac{e^{f(Q,K_i)}}{\sum_j e^{f(Q,K_j)}} \tag{3}$$

Then the soft attention is to do the weighted summation, while hard attention [17] is to choose the value with max score. In contrast, soft attention is usually widely used, and the high dimensional representation C is:

$$C = \sum_i a_i h_i \tag{4}$$

Where h_i is the i-th vector of values.

Generally key equals to value in some machine learning tasks, when query equals to key and equals to value, it's called self-attention. When different linear transformations are applied to query, key and value and performed multiple times, the attention mechanism is multi-head attention. In order to find the attention of real-time data to historical data and get a better representation for environment classification task, the basic form of soft attention is adopted in this paper.

3.2 Network Architecture

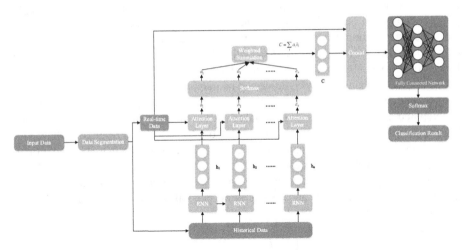

Fig. 4. Network architecture

The network architecture used in this paper is shown Fig. 4. And it mainly contains four parts, which are data segmentation, recurrent neural network, attention layers, fully connected classifier. Firstly, the input data is divided into real-time data and historical data by data segmentation. Specifically, the input data is a 257×11 matrix and the historical data is a 256×11 matrix and real-time data is a 11-demension vector. Then the historical data mentioned above are fed into the RNN, and the specific RNN used here is GRU, which structure is shown in Fig. 5:

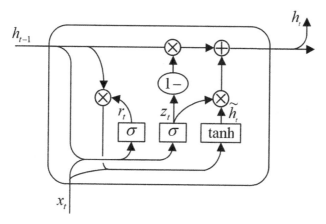

Fig. 5. The structure of gated recurrent unit (GRU)

It uses a reset gate r_t and a update gate z_t to control the hidden state h_t of recurrent neural network based on the previous hidden state h_{t-1} and the input x_t. The specific update strategy is as follows:

$$
\begin{aligned}
r_t &= \sigma\left(W_r \cdot \left[h_{t-1}, x_t\right]\right) \\
z_t &= \sigma\left(W_z \cdot \left[h_{t-1}, x_t\right]\right) \\
\tilde{h}_t &= \tanh\left(W \cdot \left[r_t \odot h_{t-1}, x_t\right]\right) \\
h_t &= (1 - z_t) \odot h_{t-1} + z_t \odot \tilde{h}_t
\end{aligned}
\tag{5}
$$

Where $\sigma(\cdot)$ is the sigmoid function, $\odot$ is Hadamard Product.

Attention values are then calculated using real-time characteristics and the RNN output in each time steps. And the RNN output in i-th time step h_i and the real-time data r are used to calculate the attention value $\hat{a}_i$ as follow:

$$
\hat{a}_i = \sigma(W_a[r; h_i] + b)
\tag{6}
$$

Where $\sigma(\cdot)$ is the sigmoid function, r is the vector of real-time data, W_a and b are the parameters of attention layer.

Softmax function is used to calculate the weight a_i of h_i, based on which high dimensional C features are constructed by soft attention. Then the real-time data r concatenates with the high dimensional representation C as the input of fully connected classifier. Batch normalization [18] and dropout [19] are used in the fully connected classifier to improve its generalization performance. Cross entropy is selected as the loss function which is expressed as follow:

$$
L(\theta) = -\hat{\mathbf{y}}^T \log f_\theta(\mathbf{x})
\tag{7}
$$

Where $\mathbf{x}$ is the input data, $\hat{\mathbf{y}}$ is the corresponding label and θ is the parameters of the whole network.

Then the network is trained using Adam [20] as the optimization method. The network architecture proposed in this paper is an end-to-end training model, which leads machine to mine relevant knowledge from collected data, rather than relying on some prior knowledge of human.

4 Experiment

The performance of machine learning method is greatly affected by training data. Cross validation is usually used as an objective method to evaluate the performance of models. Then this paper conducted two experiments to verify the effectiveness of the proposed method. One is a comparative experiment among LSTM, attention-based LSTM, GRU and attention-based GRU. The other is a visualization of environment classification result of some latest data collected by the same signal acquisition vehicle which are not in original data set.

The ratio of training set to testing set used in this experiment is 0.8, 10-fold cross validation is selected to evaluate the performance of models. And the training set and testing set of each fold are same to each model. In addition to the attention layer related parts, the network architecture and other hyper-parameters are same. The average training accuracy and average testing accuracy of each model are shown in Table 2, the details of comparative experiment is shown in Fig. 6. It can be clearly found that the performance of attention-based method is significantly better than typical one. And the average precision and recall for each category of the Attention-based GRU in 10-fold cross validation are shown in Table 3.

Table 2. 10-fold cross validation result of different models.

Models	Average training accuracy	Average testing accuracy
LSTM	81.5%	74.3%
GRU	93.9%	90.7%
Attention-based LSTM	93.5%	87.8%
Attention-based GRU	**97.3%**	**94.8%**

Table 3. Average precision and recall for each category of Attention-based GRU

Categories	Average precision	Average recall
Open area	90.7%	94.5%
Urban canyon	88.9%	96.6%
Tree shade	98.1%	90.1%

To visualize the classification results more intuitively, we map the results into RGB space as follows:

$$R = 255 * P(y = \text{tree shade}|\mathbf{x}, \theta)$$
$$G = 255 * P(y = \text{urban canyon}|\mathbf{x}, \theta) \qquad (8)$$
$$B = 255 * P(y = \text{open area}|\mathbf{x}, \theta)$$

Then we apply the trained model to some newly collected data in shanghai and compare the result of same route at different time. As Fig. 7 shows, the model performs well in

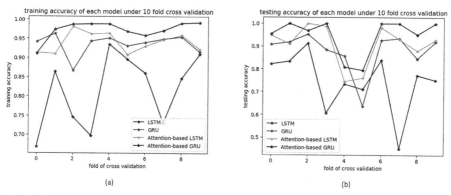

Fig. 6. The training accuracy and testing accuracy of each model under 10-fold cross validation, (a) is the figure of training accuracy, (b) is the figure of testing accuracy.

real environment. At different instants of time, the classification result of the same urban canyon (People's Square) is shown in Fig. 8. It can be found that the time may affect the model performance.

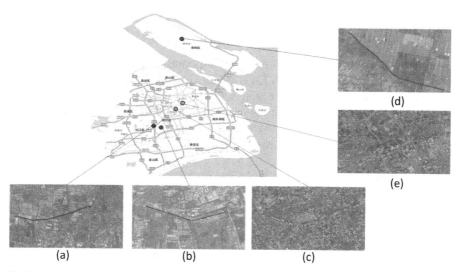

Fig. 7. An overview of the visualization of classification results (a) is an open area in Songjiang district, (b) is a tree shade in Minhang district, (c) is an urban canyon in Xuhui district, (d) is a tree shade in Chongming island, (e) is an urban canyon in Pudong district

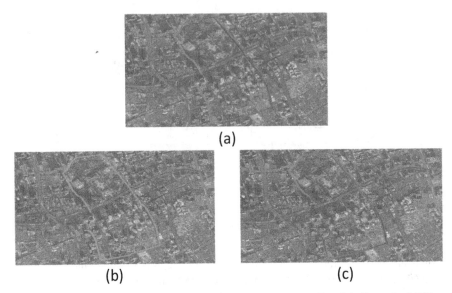

Fig. 8. The visualization of classification results at urban canyon (People's Square) of different times (a) is the result of morning, (b) is the result of afternoon, (c) is the result of night

5 Conclusion

This paper proposes an attention-based RNN to solve the problem of three typical environment classification task using NMEA-0183 data collected in real environment. The average accuracy is almost 94% on testing set of 10-fold cross validation and has significantly improvement compared to the result of RNN without attention mechanism. And the experiment of visualizing the classification results shows that the performance of proposed method is good in most cases, and the time of data collected may affect the performance a lot.

Unfortunately, our results are lack of some professional explanatory works which may extract some specific performance indicators and the generalization performance on different receivers is uncertain. Notwithstanding its limitation, this study does suggest the effectiveness of deep learning methods in environment classification task. These problems could be solved if we consider using some intermediate frequency signal characteristics or combining some transfer learning methods.

Acknowledgement. This work is supported by science and technology project of State Grid Corporation of China (No. SGSHJX00KXJS1901531).

References

1. Yanbang, J., Caixia, Z.: Development of GNSS and Its application. Value Eng. **12**, 214–215 (2013)

2. Chong, C.: The latest trends and development trends of global satellite navigation systems. Satell. Netw. **04**, 22–25 (2010)
3. Xiao, H., Wei, G., Benyu, L.: Research on GNSS Navigation Technology. GNSS World of China **3**, 59–62 (2009)
4. Misra, P., Enge, P.: Global Positioning System: Signals, Measurements, and Performance, 2nd edn. Ganga-Jamuna Press, Lincoln (2006)
5. Chao, R., Jikun, O., Yunbin, Y.: Application of adaptive filtering by selecting the parameter weight factor in precise kinematic GPS positioning. Prog. Nat. Sci. **15**(1), 41–46 (2005)
6. Tang, I., Breckon, T.P.: Automatic road environment classification. IEEE Trans. Intell. Transp. Syst. **12**(2), 476–484 (2011)
7. Gurtner W, Estey L.: RINEX - the receiver independent exchange format - Version 3.00, Astronomical Institute, University of Bern and UNAVCO, Boulder, CO (2007)
8. Gao, H., Groves, P.D.: Environmental context detection for adaptive navigation using GNSS measurements from a smartphone. Navigation **65**, 99–116 (2007)
9. Hsu, L.T., Gu, Y., Kamijo, S.: Intelligent viaduct recognition and driving altitude determination using GPS data. IEEE Trans Intell Veh **2**(3), 175–184 (2017)
10. Lighari, R.U.R., Berg, M., Salonen, E.T., et al.: Classification of GNSS SNR data for different environments and satellite orbital information. In: European Conference on Antennas & Propagation, Paris, France. IEEE (2017)
11. Munin, E., Blais, A., Couellan, N.: Convolutional neural network for multipath detection in GNSS receivers. arXiv preprint arXiv:1911.02347 (2019)
12. Yuze, W., Peilin, L., Qiang, L., et al.: Urban environment recognition based on the GNSS signal characteristics. J. Znstitute of Navig. 1–15 (2019)
13. Hochreiter, S., Schmidhuber, J.: Long short-term memory. Neural Comput. **9**(8), 1735–1780 (1997)
14. Kyunghyun, C., Bart van, B., Caglar, G., et al.: Learning phrase representations using RNN encoder-decoder for statistical machine translation. arXiv preprint arXiv:1406.1078. EMNLP 2014 (2014)
15. Bahdanau, D., Cho, K., Bengio, Y.: Neural machine translation by jointly learning to align and translate. arXiv preprint arXiv:1409.0473 (2014)
16. Vaswani, A., Shazeer, N., Parmar, N., et al.: Attention is all you need. In: 31st Conference on Neural Information Processing Systems (NIPS 2017), Long Beach, CA, USA (2017)
17. Anderson, P., et al.: Bottom-up and top-down attention for image captioning and visual question answering. arXiv preprint arXiv:1707.07998 (2017)
18. Ioffe, S., Szegedy, C.: Batch normalization: accelerating deep network training by reducing internal covariate shift. International Conference on International Conference on Machine Learning, pp. 448–456. JMLR.org (2015)
19. Hinton, G.E., Srivastava, N., Krizhevsky, A., et al.: Improving neural networks by preventing co-adaptation of feature detectors. Comput. Sci. **3**(4), 212–223 (2012)
20. Kingma, D.P., Ba, J.: Adam: a method for stochastic optimization. arXiv preprint arXiv:1412.6980 (2014)

Fast Training of POI Recommendation Models Using Gradient Compression

Hao Sun, Yunzhuo Wang, Jingwei Sun$^{(\boxtimes)}$, and Guangzhong Sun

University of Science and Technology of China, Hefei, Anhui 230026, China
{ustcsh,wyz1995}@mail.ustc.edu.cn, {sunjw,gzsun}@ustc.edu.cn

Abstract. Point-of-Interest (POI) recommendation plays an important role in location-based services. Regarding POI recommendation as a trajectory prediction problem, attention-based models have shown promising achievements to tackle this problem. With the rapid development of LBSN applications, modern training datasets and POI recommendation models are becoming increasingly large and complicated, which results in time-consuming training procedures. Existing solutions adopt data parallel training for POI models, but they still suffer from the high latency of inter-device data communication. To address this issue, we propose a fast training method for POI recommendation models. It conducts a layer-wise top-k gradient compression to reduce the overhead of data transfer. Meanwhile, it modifies the Adam optimizer with an error feedback mechanism to avoid the decrease of POI recommendation accuracy resulting from gradient compression. Experimental results show that our methods can achieve similar even better accuracy than uncompressed methods while the communication time is shorter.

Keywords: Point of interest recommendation · Gradient compression · Data parallel training

1 Introduction

In recent years, sharing experience of points of interest (POI), like restaurants and shopping malls, on location-based social networks (LBSN) has been becoming popular. It generates abundant check-in data, which enables predicting human mobility and making a recommendation of POI to users. Personalized recommendation of POI plays an important role in LBSN applications for both the users and the service providers, since it can facilitate location-based services related to dining, transportation, entertainment, etc.

POI recommendation is a problem that predicts and ranks a user's next POIs with the user's history check-ins trajectory and current GPS coordinates. It has attracted many researchers' attention in developing models. Collaborated Filter (CF) based models [7,13,20,35] and Recurrent Neural Networks (RNN) based models [10,23,29] are proposed to model users' trajectories. In recent years, attention-based models [31] are well-developed in the Natural Language

© Springer Nature Switzerland AG 2021
X. Meng et al. (Eds.): SpatialDI 2020, LNCS 12567, pp. 72–86, 2021.
https://doi.org/10.1007/978-3-030-69873-7_6

Processing (NLP) domain. It has been proved that attention-based POI recommendation models [19] can achieve state-of-the-art performance.

With the rapid development of LBSN applications, modern training datasets and POI recommendation models are becoming increasingly large and complicated. For instance, Foursquare[1] has seen over 13 billion check-ins. Such large datasets and models result in long training time for POI recommendation. Synchronous data parallel training is a conventional solution to accelerate the training. It equally separates a batch of input into several smaller batches, distributes them to multiple computation devices and computes updates to the model in parallel. Ideally, the performance of the training can scale linearly with respect to the number of separated batches. However, in practice, data parallel training requires inter-device communications of gradients to compute consistent updating directions, which prevents the performance of training from scaling linearly.

To reduce the overhead of inter-device data transfer and further improve the performance of data parallel training, gradient compression methods are proposed. When training deep neural networks using stochastic gradient descent, accurate gradients are not necessary. Therefore, lossy compression on gradients can be adopted before sending the gradients to other devices in data parallel training.

Gradient compression can effectively reduce the size of data to send so that we can train a model in a shorter time. However, it is still challenging to apply gradient compression to train deep neural networks, especially for attention-based POI recommendation models. One challenge is how to leverage the structural characterization of typical POI recommendation models to design an effective gradient compression method. The other challenge is that Adam optimizer, which is the most frequently-used optimizer for training POI recommendation models, cannot adapt to the gradient compression. Thus directly using Adam to train the model with gradient compression will diminish the recommendation accuracy.

To address these problems when applying gradient compression to train POI recommendation models, we propose an Adaptive Topk (ATopk) method to compress gradients. It eliminates redundant elements of gradients of the embedding layers and adaptively chooses the largest k elements according to the ratio of the gradient value and the parameter value. Besides, we propose an Error Feedback Nesterov (EFN) optimizer to update the model parameter from the accumulated gradient. It takes the residual error of each compression operation on gradient into account, so that the negative impact of gradient compression on model accuracy can be alleviated.

To summarize, our contributions are listed as follows.

- We propose a layer-wise adaptive Topk gradient compression method for training POI recommendation models. It achieves better recommendation accuracy than the traditional Topk method while they have similar communication performance.
- We present an error feedback Nesterov optimizer to deal with the accuracy decrease caused by gradient compression. Compared with existing Nesterov

---

[1] https://foursquare.com/about.

and Adam optimizers, it can effectively improve the model accuracy when using gradient compression.
- We conduct a detailed empirical evaluation to compare the proposed methods and existing studies on public datasets. We also make an insightful analysis and the results validate the effectiveness of our methods.

The rest of this paper is organized as follows. Section 2 reviews a series of existing studies related to this paper. Section 3 describes the framework of the proposed method. Section 4 presents the experimental results. Section 5 summarizes this paper and discusses our plan for future studies.

2   Related Work

In this section, we first review modern POI recommendation models. Then we introduce existing gradient compression methods.

2.1   POI Recommendation

Location recommendation has been attracting researchers' attention for a long period of time. Related studies evolve from Collaborated Filter (CF) [7,13,20,35] to deep learning methods. Cheng and Lian et al. [8,21] apply tensor factorization method for pairwise interaction. He et al. [14] model users' latent behavior pattern using a third-rank tensor. As an important component of deep POI recommendation models, embedding learning has attracted extra attention. Feng et al. [12] propose a personalized metric embedding A represent learning method based on the word2vec framework is developed in [11]. Some works combine traditional CF models and deep learning models. Yin et al. [36] jointly perform deep representation learning for POIs from heterogeneous features and hierarchically additive representation learning for spatial-aware personal preferences. Cheng et al. [9] propose a dual-embedding based deep latent factor model. In addition to learning a single embedding for a user, it represents each user with an additional embedding from the perspective of the interacted items.

Recurrent Neural Networks (RNN) have been widely adopted for POI recommendation due to their high accuracy of modeling sequential data. Liu et al. [23] propose a Spatial Temporal Recurrent Neural Network. It models local temporal and spatial contexts in each layer with time-specific transition matrices for different time intervals and distance-specific transition matrices for different geographical distances. Manotumruksa et al. [25] propose a novel Contextual Attention Recurrent Architecture (CARA) that can leverage both sequences of feedback and contextual information associated with the sequences to capture the users' dynamic preferences. Feng et al. [10] design a multi-modal embedding recurrent neural network to capture the complicated sequential transitions by jointly embedding the multiple factors that govern the human mobility. Zhao et al. [37] propose a new Spatio-Temporal Gated Network (STGN) by enhancing long-short term memory network, where spatio-temporal gates are introduced to capture the spatio-temporal relationships between successive check-ins. Sun et al. [29] propose a model consisting of a nonlocal network for long-term preference modeling and a geo-dilated RNN for short-term preference learning.

The attention-based model is an emerging type of solution in recent years, which can achieve superior performance for sequence modeling problems, especially for NLP tasks [31]. Li et al. [18] employs a LSTM-based encoder-decoder framework, which is able to automatically learn deep spatial-temporal representations for historical check-in activities and integrate multiple contextual factors using the embedding method in a unified manner. Lian et al. [19] propose a self-attention network called GeoSAN. It represents the hierarchical gridding of each GPS point with a self-attention based geography encoder, making better use of geographical information. Chang et al. [5] leverage a check-in context layer to capture the geographical influence and a text content layer to capture the characteristics of POIs from the text content.

2.2 Gradient Compression

Gradient compression aims to reduce communication data size, which is the main scaling bottleneck in the data parallel training. Existing studies of gradient compression can be classified into two categories. The first is compressing the gradient information using quantization or sparsification. The second is how to utilize the compressed gradient to update parameters.

Gradient Quantization. Gradient quantization represents the gradient with a low-precision format. Seide et al. [26] propose a 1-bit SGD method to reduce communication. QSGD method [2] is a family of compression schemes for gradient updates, which allows the user to smoothly trade-off communication bandwidth and convergence time and provides convergence guarantees. TernGrad [34] uses ternary gradients, which requires only three numerical levels $-1,0,1$, to accelerate distributed deep learning in data parallelism. SignSGD [4] compresses the gradient to the sign(x) function. It represents the major vote of the gradient aspects, which only takes one bit for each element.

Gradient Sparsification. Aji et al. [1] and Strom [28] propose gradient sparsification methods that only transfer the elements of gradients within a threshold. Stich et al. [27] propose a top-k method that chooses the k largest elements to represent the whole gradient. Wangni et al. [33] propose a unbiased random sparsification method that sets an element x to 0 with probability $1 - p$ and $\frac{x}{p}$ with probability p.

Updating Parameter with Compressed Gradient. Lin et al. [22] point out that 99.9% gradient exchange is redundant, and propose a momentum correction method called Deep Gradient Compression. Stich et al. [27] prove that SGD top-k or random-k converges at the same rate as vanilla SGD when equipped with

---

Algorithm 1: Adaptive TopK gradient compression method

Input: parameters $param$, gradients g, compression parameter k
Output: compressed gradient list $ListG$, index list $ListIndex$
1 $ListG = \emptyset, ListIndex = \emptyset$;
2 **for** $gradient\ g[layer\_name]\ layer\_name \in model$ **do**
3   num_elements = $\text{len}(g[layer\_name])$;
4   **if** $layer\_name\ is\ location\_embedding$ **then**
5   **for** $i=0,1,..,num\_elements\text{-}1$ **do**
6   **if** $g[layer\_name][i]\ is\ not\ 0$ **then**
7               $ListG.\text{append}(g[layer\_name][i])$;
8               $ListIndex.\text{append}(layer\_name,i)$;
9   **end**
10   **end**
11   **end**
12   **else**
13   **for** $i=0,1,..,num\_elements\text{-}1$ **do**
14   **if** $\frac{|g[layer\_name][i]|}{|param[layer\_name][i]|+\epsilon)}$ $is\ the\ largest\ k\times\ num\_elements$ **then**
15              $ListG.\text{append}(g[layer\_name][i])$;
16              $ListIndex.\text{append}(layer\_name,i)$;
17   **end**
18   **end**
19   **end**
20 **end**

---

error compensation. DoubleSquezee [30] gives a theoretical convergence analysis on error-compensated SGD method. DORE [24] does not require bound gradient, which is different from DoubleSquezee, but it assumes that the compression operator is unbiased. Karimireddy et al. [17] use error feedback to correct the biased sign compression operator. AdaComp [6] is based on a simple local sampling strategy to adapt the compression rate. Vogels et al. [32] propose PowerSGD, a gradient compression operator based on the low-rank approximation.

3   Method

3.1   Preliminary

Given an user u, her or his history trajectory is composed of a sequence of n check-ins $c_1^u \rightarrow c_2^u \rightarrow ... \rightarrow c_n^u$. Each check-in c_i consists of (u_i, l_i, p_i) where p_i is a GPS coordinate containing a latitude and a longitude. In this paper, we regard POI recommendation as a problem that given user u's history trajectory $c_1^u \rightarrow c_2^u \rightarrow ... \rightarrow c_n^u$ and current GPS p_{n+1} to recommend next location l_{n+1}. We take GeoSAN [19], a state-of-the-art POI recommendation model, as an instance to demystify the overhead of communicating gradient in training a POI recommendation model with data parallel scheme.

3.2 Adaptive TopK Gradient Compression

Data parallelism is a standard approach to accelerate the training of deep neural networks. In this approach, a number of computing devices, usually GPUs in typical scenarios, concurrently fetch different batches of input data and maintain the same copies of the model via exchanging local gradient information. The computation on the devices works independently to each other, thus it can well scale. In contrast, the communication for maintaining model consistency must involve all devices and cause additional overhead, so that it usually becomes a bottleneck of the Data parallel training. To reduce this overhead, gradient compression methods are proposed. Topk SGD is one of the most useful SGD variants with a gradient compression mechanism.

In the traditional Topk method, each device communicates the k largest elements of the gradient to approximate the whole gradient. Being aware of the size of different layers can vary dramatically, Topk method usually sets different sparsity parameter k according to some percentage of the size of each layer. However, the traditional Topk method only considers the absolute values of the gradients as their importance, but not takes the relative impact of the gradient on the parameter into account. In other words, it is not sufficient to use only the absolute values of the gradients to measure whether an element should be eliminated.

To address this issue, we propose an adaptive Topk gradient compression method. The detailed procedure of the adaptive Topk gradient compression is illustrated by Algorithm 1. It consists of two stages to update the parameters. The first is that we send all nonzero gradients in the location embedding layers, which has large amounts of gradients but only a small portion of them needs to be updated. According to our experiment, we find that most elements of location embedding layers are nonzero. The second is that before calculating the top-k elements of each layer, we normalize the value of each element in the gradients using the corresponding value of parameters. Such a process results in that the following top-k operation will adaptively pick up the gradient elements that move k largest relative step size over their corresponding parameters.

3.3 Distributed Error Feedback for Nesterov Optimizer

Straightforwardly applying sparsifed gradient to the vanilla SGD without additional processing will not achieve a promising accuracy because of losing gradient information. When we apply ATopk on training the POI recommendation model, the lossy gradient also diminishes the recommendation accuracy.

To tackle this problem, researchers [3,27] propose an error feedback method to recover the lost information of the sparsified gradient. The method stores the residual value between the gradient value and the compressed value in local memory. Inspired by this method, we design an optimizer called Error Feedback Nesterov (EFN), based on the existing Nesterov optimizer. The Nesterov optimizer considers the future influence on the gradient, which can be regarded as the second-order derivative information. We further involve second-order momentum

---

Algorithm 2: Training with EFN optimizer

Input: learning rate γ, first momentum parameter m_1, second momentum
        parameter m_2, model parameter x, momentum $m = 0$, workers
        $w = \{1, ..., W\}$

Output: updated gradient x'

1 **for** *each worker $w = 1, ..., W$* **do**
2 | memory $e_w = 0$;
3 | **for** *each iterate $t = 0, ...$* **do**
4 | | Compute a stochastic gradient g_w;
5 | | $\Delta_w = g_w + e_w$;
6 | | $C(\Delta_w)$ =COMPRESS(Δ_w);
7 | | $C(\Delta)$ =AGGREGATE($C(\Delta_1), ..., C(\Delta_w)$);
8 | | $e_w = \Delta_w - C(\Delta_w)$;
9 | | $m_1 = \lambda_1 m_1 + C(\Delta)$;
10 | | $m_2 = \lambda_2 m_2 + C(\Delta) * C(\Delta)$;
11 | | $x = x - \gamma * (x + m_1)/(\sqrt{x * x + m_2} + 1e - 8)$
12 | **end**
13 **end**

---

information to Nesterov, and integrate the error feedback mechanism into the
modified Nesterov optimizer.

The proposed EFN optimizer is implemented as Algorithm 2 shows. First,
the optimizer corrects the gradient g_w from the error feedback e_w, where w
denotes a worker, namely a computing device. The corrected gradient is $\Delta_w = g_w + e_w$. Then EFN compresses the gradients through a compression function.
The compressed gradient is

$$C(\Delta_w) = COMPRESS(\Delta_w)$$

Each worker communicates with other workers and aggregates the gradients.
After aggregating, the compressed gradients are changed to

$$C(\Delta) = AGGREGATE(C(\Delta_1), ..., C(\Delta_w))$$

Then EFN stores the error feedback $e_w = \Delta_w - C(\Delta_w)$.

In the following steps, the optimizer calculates the first-order and the second-
order momentum, respectively, and then updates the model parameter. The first-
order momentum is calculated by $m_1 = \lambda_1 m_1 + C(\Delta)$, while the second-order
momentum is $m_2 = \lambda_2 m_2 + C(\Delta) * C(\Delta)$. Last, the parameter is updated by

$$x = x - \gamma * (x + m_1)/(\sqrt{x * x + m_2} + 1e - 8)$$

4 Experiment

4.1 Experiment Setup

Environment. Our platform is a server that is equipped with $2 \times$ Intel Xeon CPUs and $8 \times$ Nvidia 2080ti GPUs. The operating system is CentOS. Our experiments run under the framework of Pytorch 1.6.0.

Dataset. We evaluate our method on two widely used LBSN datasets: Gowalla and Brightkite. We remove user data with less than 20 check-ins and locations that are visited fewer than 10 times. Table 1 shows a summary of the two datasets. For the check-in sequence of each user, we divide it into non-overlapping subsequences of length 100 and the latest 100 check-ins will be used in the evaluation.

Table 1. Statistics of the two datasets.

	Brightkite	Gowalla
# user	5,247	31,708
# locations	48,181	131,329
# check-ins	1,699,579	2,963,373

Baseline. We compare our model with three representative models to show its effectiveness.

- GRU is a simple baseline model based on GRU4Rec [15], which adopts a single-layer GRU for modeling sequence data.
- STGN [37] improves LSTM with two pairs of spatio and temporal gates, which can model both short-term interest and long-term interest simultaneously.
- GeoSAN [19] constructs an encoder-decoder model, which uses a self-attention mechanism to express the correlation among sequence data. In the implementation of GeoSAN, we adopt a uniform sampler to get negative samples and weighted binary cross-entropy loss function in [19] for model training.

Metrics. We adopt widely used metrics to evaluate performance: Hit Rate. Hit Rate at a cutoff k, denoted as HR@k, counts the rate of correct recommendations where the target location is correctly contained in the top k recommendations. In our experiments, we set $k = \{5,10\}$ and we choose a uniform sampler to get negative samples.

4.2 Matching Performance

The performance of GeoSAN with ATopk strategy and EFN optimizer and the three baselines on two datasets, evaluated by Hit@5 and Hit@10, is shown in Table 2. GeoSAN reaches the best result in the three baselines. Our proposed methods achieve the best results on two public datasets with two metrics. It proves that our ATopk compression method and EFN optimizer can get promising matching performance when training the GeoSAN for POI recommendation.

The cell size is equal to sequence length and the hidden layer size is set to 128 in our experiments. The number of epochs is set to 50 and the batch size is set to 64. In this paper, we choose $k = \{5,10\}$ to show different results of Hit@k. We use the Adam optimizer when we training the GRU, STGN, and GeoSAN model. For GRU, STGN, and GeoSAN, we choose the result of the 10th epoch as the final result, because at the 10th epoch we observe that the model has converged. For GeoSAN+ATopk+EFN, we choose the last epoch as the final result. For Brightkite, we set the compression rate to 0.05 and for Gowalla we set that to 0.03.

Table 2. Matching performance.

	Brightkite		Gowalla	
	Hit@5	Hit@10	Hit@5	Hit@10
GRU	0.2680	0.3665	0.4025	0.5436
STGN	0.2175	0.2979	0.3932	0.5320
GeoSAN	0.3415	0.4679	0.5218	0.6654
GeoSAN+ATopk+EFN	0.437691	0.572790	0.576479	0.713952

Table 3. Performance of average model on Brightkite.

	Brightkite			
	Hit@5	AveHit@5	Hit@10	AveHit@10
ATopk+EFN	0.433308	0.407965	0.713952	0.543445
ATopk+Nesterov	0.334604	0.305259	0.463986	0.430069
Topk+Adam	0.321456	0.358041	0.450838	0.491616

4.3 Average Model

Inspired by Izmailov et al. [16], we average the history model parameters with the same weight and test the average model. In this experiment, we show Hit@5 and Hit@10 of a model and average model under different combinations of compression strategies and optimizers, respectively. The results of the average model are denoted as AveHit@5 and AveHit@10. On Brightkite, we set 4 layers, 8 heads

Table 4. Performance of average model on Gowalla.

	Gowalla			
	Hit@5	AveHit@5	Hit@10	AveHit@10
ATopk+EFN	0.578182	0.540652	0.718809	0.685694
ATopk+Nesterov	0.580074	0.538129	0.715466	0.679387
Topk+Adam	0.478901	0.476315	0.629053	0.628863

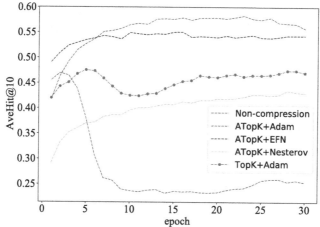

Fig. 1. AveHit@10 versus number of epochs on the Brightkite.

self-attention module for the encoder, and 1 layer, 8 heads self-attention module for the decoder. The compression rate is 0.05. On Gowalla, we set 2 layers, 2 heads self-attention module for the encoder and 1 layer, 2 heads self-attention module for the decoder. The compression rate is 0.03. Through observation, we found that the model tends to converge in about 30 epochs, so we choose the last epoch to compare Hit@10 and AveHit@10.

4.4 HitRate with Different Training Settings

In this section, we show the HitRate of POI recommendation on testing datasets, with different training settings. The experimental configurations of the number of epochs and hyper-parameters are consistent with that in Sect. 4.3, thus these experiments are also based on average models and the results differ from that in Sect. 4.2. We compare the performance of different combinations between optimizers and gradient compression strategies, as Fig. 1 and Fig. 2 illustrate.

Broadly speaking, on the Brightkite dataset, ATopK+EFN achieves the best performance, compared with other combinations of compression strategies and optimizers. Besides achieving a fast convergence rate, it is very close to the uncompressed case, which is the upper bound of the model accuracy in this

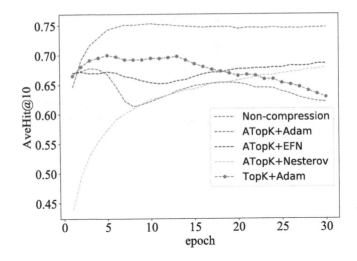

Fig. 2. AveHit@10 versus number of epochs on the Gowalla.

comparison. The experimental results on the Gowalla dataset are also consistent with that on the Brightkite dataset. The results validate that ATopK+EFN can express gradient information well and this combination is also compatible with model averaging. Meanwhile, with the ATopK strategy, the optimizing result of EFN is better than that of Nesterov and Adam. It indicates that with our compression strategy, it is effective to involve the second-order momentum information and the second-order derivative information for updating parameters.

4.5   Communication

We evaluate the communication times on the Brighkite dataset using ATopk, Topk and non-compression method with {2, 4, 8} workers, respectively. The blue and yellow areas in Fig. 3 represent the average communication time consumption of the ATopk and Topk method in one training iteration, respectively. The green area represents the time that the non-compression method communication takes.

As Fig. 3 illustrates, communication time of the ATopk and Topk method is significantly smaller than that of the non-compression method. The difference between ATopk and Topk is relatively small. Compared to the non-compression, the ATopk method reduces 90.71%, 69.02%, 35.99% time for different numbers of workers.

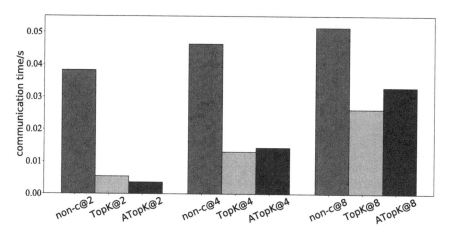

Fig. 3. Communication. (Color figure online)

5 Conclusion

In this paper, we propose ATopk compression method and EFN optimizer for the sparesifed gradient update in training POI recommendation models. ATopk compression can effectively reduce the communication data size of data parallel training of POI recommendation models, thus the traning time cost can be saved. To eliminate the negative impact from the compression on model accuracy, EFN optimizer can be adopted to accumulate and feedback the residual errors generated by ATopk. Experiments shows that our methods can achieve promising model performance for training GeoSAN model on Brightkite and Gowalla. Compared with existing studies, the proposed methods can get similar even better accuracy, while the communication time cost is reduced.

Although we have achieved desired improvement in the experiments that well represent typical POI recommendation applications, the ability of our methods can be further optimized. In the future, we will further investigate the convex analysis on ATopk method and EFN, and explore potential approaches to scale the training to larger models and datasets.

References

1. Aji, A.F., Heafield, K.: Sparse communication for distributed gradient descent. In: Proceedings of the 2017 Conference on Empirical Methods in Natural Language Processing, EMNLP 2017, Copenhagen, Denmark, 9–11 September 2017, pp. 440–445 (2017)
2. Alistarh, D., Grubic, D., Li, J., Tomioka, R., Vojnovic, M.: QSGD: communication-efficient SGD via gradient quantization and encoding. In: Advances in Neural Information Processing Systems 30: Annual Conference on Neural Information Processing Systems 2017, 4–9 December 2017, Long Beach, CA, USA, pp. 1709–1720 (2017)

3. Alistarh, D., Hoefler, T., Johansson, M., Konstantinov, N., Khirirat, S., Renggli, C.: The convergence of sparsified gradient methods. In: Advances in Neural Information Processing Systems 31: Annual Conference on Neural Information Processing Systems 2018, NeurIPS 2018, 3–8 December 2018, Montréal, Canada, pp. 5977–5987 (2018)

4. Bernstein, J., Wang, Y., Azizzadenesheli, K., Anandkumar, A.: SIGNSGD: compressed optimisation for non-convex problems. In: Proceedings of the 35th International Conference on Machine Learning, ICML 2018, Stockholmsmässan, Stockholm, Sweden, 10–15 July 2018, pp. 559–568 (2018)

5. Chang, B., Park, Y., Park, D., Kim, S., Kang, J.: Content-aware hierarchical point-of-interest embedding model for successive POI recommendation. In: Proceedings of the Twenty-Seventh International Joint Conference on Artificial Intelligence, IJCAI 2018, 13–19 July 2018, Stockholm, Sweden, pp. 3301–3307 (2018)

6. Chen, C., Choi, J., Brand, D., Agrawal, A., Zhang, W., Gopalakrishnan, K.: Adacomp: adaptive residual gradient compression for data-parallel distributed training. In: Proceedings of the Thirty-Second AAAI Conference on Artificial Intelligence, (AAAI-18), the 30th Innovative Applications of Artificial Intelligence (IAAI-18), and the 8th AAAI Symposium on Educational Advances in Artificial Intelligence (EAAI-18), New Orleans, Louisiana, USA, 2–7 February 2018, pp. 2827–2835 (2018)

7. Cheng, C., Yang, H., King, I., Lyu, M.R.: Fused matrix factorization with geographical and social influence in location-based social networks. In: Proceedings of the Twenty-Sixth AAAI Conference on Artificial Intelligence, 22–26 July 2012, Toronto, Ontario, Canada (2012)

8. Cheng, C., Yang, H., Lyu, M.R., King, I.: Where you like to go next: successive point-of-interest recommendation. In: IJCAI 2013, Proceedings of the 23rd International Joint Conference on Artificial Intelligence, Beijing, China, 3–9 August 2013, pp. 2605–2611 (2013)

9. Cheng, W., Shen, Y., Zhu, Y., Huang, L.: DELF: a dual-embedding based deep latent factor model for recommendation. In: Proceedings of the Twenty-Seventh International Joint Conference on Artificial Intelligence, IJCAI 2018, 13–19 July 2018, Stockholm, Sweden, pp. 3329–3335 (2018)

10. Feng, J., et al.: DeepMove: predicting human mobility with attentional recurrent networks. In: Proceedings of the 2018 World Wide Web Conference on World Wide Web, WWW 2018, Lyon, France, 23–27 April 2018, pp. 1459–1468 (2018)

11. Feng, S., Cong, G., An, B., Chee, Y.M.: PoI2VEC: geographical latent representation for predicting future visitors. In: Proceedings of the Thirty-First AAAI Conference on Artificial Intelligence, 4–9 February 2017, San Francisco, California, USA, pp. 102–108 (2017)

12. Feng, S., Li, X., Zeng, Y., Cong, G., Chee, Y.M., Yuan, Q.: Personalized ranking metric embedding for next new POI recommendation. In: Proceedings of the Twenty-Fourth International Joint Conference on Artificial Intelligence, IJCAI 2015, Buenos Aires, Argentina, 25–31 July 2015, pp. 2069–2075 (2015)

13. Gao, H., Tang, J., Hu, X., Liu, H.: Content-aware point of interest recommendation on location-based social networks. In: Proceedings of the Twenty-Ninth AAAI Conference on Artificial Intelligence, 25–30 January 2015, Austin, Texas, USA, pp. 1721–1727 (2015)

14. He, J., Li, X., Liao, L., Song, D., Cheung, W.K.: Inferring a personalized next point-of-interest recommendation model with latent behavior patterns. In: Proceedings of the Thirtieth AAAI Conference on Artificial Intelligence, 12–17 February 2016, Phoenix, Arizona, USA, pp. 137–143 (2016)

15. Hidasi, B., Karatzoglou, A., Baltrunas, L., Tikk, D.: Session-based recommendations with recurrent neural networks. In: Bengio, Y., LeCun, Y. (eds.) 4th International Conference on Learning Representations, ICLR 2016, San Juan, Puerto Rico, 2–4 May 2016. Conference Track Proceedings (2016)
16. Izmailov, P., Podoprikhin, D., Garipov, T., Vetrov, D.P., Wilson, A.G.: Averaging weights leads to wider optima and better generalization. In: Proceedings of the Thirty-Fourth Conference on Uncertainty in Artificial Intelligence, UAI 2018, Monterey, California, USA, 6–10 August 2018, pp. 876–885 (2018)
17. Karimireddy, S.P., Rebjock, Q., Stich, S.U., Jaggi, M.: Error feedback fixes sign SGD and other gradient compression schemes. In: Proceedings of the 36th International Conference on Machine Learning, ICML 2019, 9–15 June 2019, Long Beach, California, USA, pp. 3252–3261 (2019)
18. Li, R., Shen, Y., Zhu, Y.: Next point-of-interest recommendation with temporal and multi-level context attention. In: IEEE International Conference on Data Mining, ICDM 2018, Singapore, 17–20 November 2018, pp. 1110–1115 (2018)
19. Lian, D., Wu, Y., Ge, Y., Xie, X., Chen, E.: Geography-aware sequential location recommendation. In: KDD 2020: The 26th ACM SIGKDD Conference on Knowledge Discovery and Data Mining, Virtual Event, CA, USA, 23–27 August 2020, pp. 2009–2019 (2020)
20. Lian, D., Zhao, C., Xie, X., Sun, G., Chen, E., Rui, Y.: Geomf: joint geographical modeling and matrix factorization for point-of-interest recommendation. In: The 20th ACM SIGKDD International Conference on Knowledge Discovery and Data Mining, KDD 2014, New York, NY, USA, 24–27 August 2014, pp. 831–840 (2014)
21. Lian, D., Zheng, V.W., Xie, X.: Collaborative filtering meets next check-in location prediction. In: 22nd International World Wide Web Conference, WWW 2013, Rio de Janeiro, Brazil, 13–17 May 2013, Companion Volume, pp. 231–232 (2013)
22. Lin, Y., Han, S., Mao, H., Wang, Y., Dally, B.: Deep gradient compression: Reducing the communication bandwidth for distributed training. In: 6th International Conference on Learning Representations, ICLR 2018, Vancouver, BC, Canada, April 30 - May 3 2018, Conference Track Proceedings (2018)
23. Liu, Q., Wu, S., Wang, L., Tan, T.: Predicting the next location: a recurrent model with spatial and temporal contexts. In: Proceedings of the Thirtieth AAAI Conference on Artificial Intelligence, 12–17 February 2016, Phoenix, Arizona, USA, pp. 194–200 (2016)
24. Liu, X., Li, Y., Tang, J., Yan, M.: A double residual compression algorithm for efficient distributed learning. In: The 23rd International Conference on Artificial Intelligence and Statistics, AISTATS 2020, 26–28 August 2020, Palermo, Sicily, Italy, pp. 133–143 (2020)
25. Manotumruksa, J., Macdonald, C., Ounis, I.: A contextual attention recurrent architecture for context-aware venue recommendation. In: The 41st International ACM SIGIR Conference on Research & Development in Information Retrieval, SIGIR 2018, Ann Arbor, MI, USA, 08–12 July 2018, pp. 555–564 (2018)
26. Seide, F., Fu, H., Droppo, J., Li, G., Yu, D.: 1-bit stochastic gradient descent and its application to data-parallel distributed training of speech DNNs. In: INTERSPEECH 2014, 15th Annual Conference of the International Speech Communication Association, Singapore, 14–18 September 2014, pp. 1058–1062 (2014)
27. Stich, S.U., Cordonnier, J., Jaggi, M.: Sparsified SGD with memory. In: Advances in Neural Information Processing Systems 31: Annual Conference on Neural Information Processing Systems 2018, NeurIPS 2018, 3–8 December 2018, Montréal, Canada, pp. 4452–4463 (2018)

28. Strom, N.: Scalable distributed DNN training using commodity GPU cloud computing. In: INTERSPEECH 2015, 16th Annual Conference of the International Speech Communication Association, Dresden, Germany, 6–10 September 2015, pp. 1488–1492 (2015)

29. Sun, K., Qian, T., Chen, T., Liang, Y., Nguyen, Q.V.H., Yin, H.: Where to go next: Modeling long- and short-term user preferences for point-of-interest recommendation. In: The Thirty-Fourth AAAI Conference on Artificial Intelligence, AAAI 2020, The Thirty-Second Innovative Applications of Artificial Intelligence Conference, IAAI 2020, The Tenth AAAI Symposium on Educational Advances in Artificial Intelligence, EAAI 2020, New York, NY, USA, 7–12 February 2020, pp. 214–221 (2020)

30. Tang, H., Yu, C., Lian, X., Zhang, T., Liu, J.: Doublesqueeze: parallel stochastic gradient descent with double-pass error-compensated compression. In: International Conference on Machine Learning, pp. 6155–6165. PMLR (2019)

31. Vaswani, A., et al.: Attention is all you need. In: Advances in Neural Information Processing Systems 30: Annual Conference on Neural Information Processing Systems 2017, 4–9 December 2017, Long Beach, CA, USA, pp. 5998–6008 (2017)

32. Vogels, T., Karimireddy, S.P., Jaggi, M.: PowerSGD: practical low-rank gradient compression for distributed optimization. In: Advances in Neural Information Processing Systems 32: Annual Conference on Neural Information Processing Systems 2019, NeurIPS 2019, 8–14 December 2019, Vancouver, BC, Canada, pp. 14236–14245 (2019)

33. Wangni, J., Wang, J., Liu, J., Zhang, T.: Gradient sparsification for communication-efficient distributed optimization. In: Advances in Neural Information Processing Systems 31: Annual Conference on Neural Information Processing Systems 2018, NeurIPS 2018, 3–8 December 2018, Montréal, Canada, pp. 1306–1316 (2018)

34. Wen, W., et al.: Terngrad: ternary gradients to reduce communication in distributed deep learning. In: Advances in Neural Information Processing Systems 30: Annual Conference on Neural Information Processing Systems 2017, 4–9 December 2017, Long Beach, CA, USA, pp. 1509–1519 (2017)

35. Ye, M., Yin, P., Lee, W., Lee, D.L.: Exploiting geographical influence for collaborative point-of-interest recommendation. In: Proceeding of the 34th International ACM SIGIR Conference on Research and Development in Information Retrieval, SIGIR 2011, Beijing, China, 25–29 July 2011, pp. 325–334 (2011)

36. Yin, H., Wang, W., Wang, H., Chen, L., Zhou, X.: Spatial-aware hierarchical collaborative deep learning for POI recommendation. IEEE Trans. Knowl. Data Eng. **29**(11), 2537–2551 (2017)

37. Zhao, P., et al.: Where to go next: a spatio-temporal gated network for next POI recommendation. In: The Thirty-Third AAAI Conference on Artificial Intelligence, AAAI 2019, The Thirty-First Innovative Applications of Artificial Intelligence Conference, IAAI 2019, The Ninth AAAI Symposium on Educational Advances in Artificial Intelligence, EAAI 2019, Honolulu, Hawaii, USA, 27 January - 1 February 2019, pp. 5877–5884 (2019)

Research on Vector Road Aided Inertial Navigation by Using ICCP Algorithm

Xiang Li[1]($\boxtimes$) ⓘ and Wenbing Liu[2]

[1] Information Engineering University, Zhengzhou 450001, Henan, China
ryolx13@126.com
[2] Army Logistics University of PLA, Chongqing 401311, China

Abstract. Inertial navigation system (INS) has played a huge advantage in a lot of special conditions. But its positioning error will accumulate with time, so it is difficult to work independently for a long time. And the vehicle loaded with the inertial navigation system usually drive on the road, so the high precision vector road data based on GIS can be used as an auxiliary information, which could correct INS errors by the correlation matching algorithm. Because the inertial navigation system does not have the problem of signal hopping or missing, the traditional GPS matching algorithms could not work well as usual. A vector road aided inertial navigation by using ICCP algorithm is proposed with the features of the vehicle trajectory and the corresponding matching road. Firstly, the matching data track data set and the corresponding road set, which have the similar shape features, are determined by the opening size, opening direction and the confidence interval of error. Further, the nearest projection points could be found out to constitute the skeleton frame of the ICCP algorithm. Then, the average of the shortest distance between two data sets is calculated by the distance and position relationship between the trajectory points and road points. Finally, the trajectory data set is transferred by the average shortest distance iteratively. When the trajectory error is convergent, the set of matching points is obtained, and the path matching is completed. The simulation experiments was conducted and compared with the traditional direct projection algorithm and feature weights algorithm with the trajectory data by inertial navigation system and high accuracy road network database. The simulation results certify that the proposed method generally performs better than the traditional GPS algorithms in the matching accuracy rate, mismatch rate and efficiency.

Keywords: ICCP algorithm · Vehicle trajectory data · Features · Road matching · Iterative transform

1 Introduction

Inertial navigation system (INS) is widely used in the field of navigation and positioning in special environment because of its good independence and reliability [1]. With the development of low-cost MEMS inertial sensors and other devices technology, commercial and consumer grade vehicle inertial navigation system has gradually entered

X. Meng et al. (Eds.): SpatialDI 2020, LNCS 12567, pp. 87–106, 2021.
https://doi.org/10.1007/978-3-030-69873-7_7

the common people's home [2]. Due to the continuous accumulation of INS error with time, it is difficult to work independently for a long time. GPS is usually used as an auxiliary navigation means. However, GPS signals are often affected by the refraction of high-rise buildings, the shielding of underground tunnels or the interference of complex viaducts in the cities, resulting in "GPS blind area" which leads to the decline of positioning accuracy [3]. The vehicle usually drives on the road, so the high-precision road data can be used as a new auxiliary means to correct the INS error through the matching algorithm, so as to improve the positioning accuracy and reliability of the INS and realize the vehicle positioning and navigation without satellite navigation.

At present, most of road matching algorithms are used for GPS navigation correction, mainly including geometry algorithm, topology algorithm, weight algorithm, etc. These algorithms are suitable for inertial navigation as well. The most commonly used direct projection algorithm has simple principle and good real-time performance [4]. During the process of algorithm, there is a lot of distance calculation, which is quite tedious. In the case of complex roads such as intersections and other special sections, the matching accuracy is relatively low [5, 6]. The topological matching algorithm mainly uses the spatial topological relationship of the road network and the connectivity between the roads in the Geographic Information System (GIS) to choose the corresponding roads to be matched. However, the construction of the topological network requires high quality of the road data and the integrity of the network, and the ordinary vector road data is often difficult to achieve the expected matching effect [7]. In reference [8], an algorithm based on weight factor is proposed, which makes more use of vehicle trajectory parameters to obtain the final matching results according to certain evaluation criteria. The value of weight coefficients such as driving direction, relative position relationship and distance are mostly determined based on empirical values, lacking scientific selection criteria [9, 10]. In the case of similar evaluation weights, the matching results are prone to jump and mismatching occurs frequently. Most of the above algorithms focus on the selection and determination of matching roads, and finally determines the matching results according to the projection of matching roads. Unlike GPS, there is no signal jump or discontinuity in the INS. It is obviously unable to meet the positioning requirements only through projection correction. Therefore, the above GPS matching algorithm is not suitable for INS.

There are also some advanced road network matching algorithms that often take comprehensive information into account, such as Kalman filter [11], fuzzy logic model [12], hidden Markov model [13], etc. Due to the use of advanced technology, the accuracy of these advanced road network matching algorithms is generally high. It is greatly affected by complex road network, sampling rate and other factors, the calculation cost is high and the real-time matching effect is not ideal [14].

In 1992, BESL and McKay [15], researchers of computer vision, introduced the idea of a high-level registration method based on free-form surfaces. Chen [16] and Bergevin [17] proposed the accurate registration method of point-to-plane search for the nearest point, which improved the ICP algorithm (iterative closest point). Because of its matching characteristics based on geometry, color or grid, this algorithm has attracted great attention by scholars in related fields, which has been widely used in image alignment, pattern recognition and position estimation. ICCP (iterative closest contour point) algorithm is a special case of ICP algorithm, which takes isoline as matching unit. Its principle is simple and easy to implement. Many scholars have discussed and studied the application and promotion of ICCP algorithm. In reference [18], this algorithm is first introduced into gravity matching aided navigation with the technical process and implementation method. Liu Fanming [19], Wang Zhigang [20], Yan Li [21] and others improved and extended the application of ICCP algorithm in gravity matching positioning, further expanding the application scope of the algorithm. In reference [22, 23], the geomagnetic characteristics were used as matching method to improve the positioning accuracy of inertial navigation.

At present, there are some research results of ICCP algorithm in the road matching of INS. Liurui proposed the improvement of ICCP algorithm based on the path matching of INS carrier in road network [24]. The problems and shortcomings of the algorithm are analyzed and the corresponding improvement is made. Li Xiang used ICCP algorithm in track and road network data matching as a matching method [25]. In reference [26], the ICCP algorithm is applied to the error correction of the inertial navigation of autonomous robot, and the ideal effect is achieved. And in reference [5], a method for tight integration of IMU (Inertial Measurement Unit), stereo VO (Visual Odometry) and digital map for land vehicle navigation was proposed, which effectively limits the quick drift of INS. Although all of them have achieved certain results, but none of the above literature has given a systematic and detailed implementation method which is suitable for INS road matching. The ICCP algorithm can effectively suppress the INS error, it is more suitable for fine matching under small error, so it is necessary to ensure that the road to be matched is near the INS track.

Based on these problems, this paper proposes a vector road data aided vehicle inertial navigation algorithm based on the research results of road matching. Firstly, the sets of road points to be matched would be chosen by using the opening size, opening direction and error confidence interval of the track feature; secondly, the closest contour point is found according to the normal direction of track heading and the translation transformation is calculated; then, the final matching track is obtained by iterative method, and the convergence and matching variance of target distance function are used to evaluate the matching results. Finally, the simulation experiment is carried out to demonstrate the feasibility of the algorithm in path matching.

2 Basic Principle of Algorithm

2.1 Basic Ideas

The essence of ICCP algorithm is to match multilateral arcs. To find a short arc and make it match a long arc best. Different from GPS, the information of trajectory points output by INS is non-abrupt, and there is no jump or discontinuity, so it is very similar to the road. The spatial position of the track points is not a single value function of the track point coordinates, and it is impossible to achieve matching and positioning with a single point. Therefore, it is necessary to obtain a section of driving track of the vehicle, forming a set of multiple track points which have some connections with road data. Under the condition that the heading angle and relative position of each inertial navigation track point do not change, the closest point on its isoline would be found, and the alignment and matching between the navigation track and the real track through the iteration of the closest point could be realized. Its principle is shown in Fig. 1.

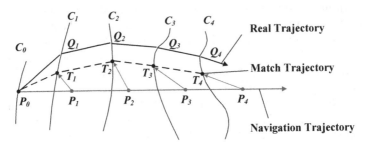

Fig. 1. The principle of trajectory matching algorithm based on ICCP

In Fig. 1, $\{P\}$ is the set of navigation track points sent by the system, $\{Q\}$ is the set of real track points corresponding to $\{P\}$, that is vector road data in map or road network database. ICCP algorithm is based on the idea of segmented matching, in order to obtain the real track of $\{P\}$, there must be some kind of rigid transformation T, which can make P_i close to Q_i and minimize the distance between the two sets, so as to find the optimal estimation point X_i and the corresponding set of matched track points $\{X\}$, and to realize the optimal estimation of the real track [27]. The rigid transformation consists of two translation components and one rotation component

$$TP = \vec{t} + RP \tag{1}$$

In formula (1), $\vec{t} = (t_x, t_y)$ represents the horizontal and vertical translation components, and R is rotation component. The process of solving the rigid transformation can be transformed into a process of finding the optimal solution. A target distance function E can be set to determine whether the estimated real trajectory points are correct. The target function E can be expressed as

$$E(X, TP) = \sum_{i=1}^{N} \|x_i - Tp_i\| = \sum_{i=1}^{N} \left\| x_i - (Rp_i + \vec{t}) \right\| \tag{2}$$

In formula (2), E represents the total distance between the set of trajectory points $\{P\}$ and the set of estimated real trajectory points $\{X\}$. The smaller the E value is, the closer the navigation trajectory is to the road data. With the convergence of T, E also reduces the convergence. When E value is the minimum, the rigid transformation t is the optimal solution. After the last iteration, the set TP is the final matching trajectory.

2.2 Key Issues

But the ICCP algorithm for INS road matching is different from gravity matching, geomagnetic matching and terrain matching. There are 3 key problems as follows:

(1) Simplification of rigid transformation. The positioning error of INS can be divided into translation error and rotation error, and these errors will drift with the increase of time and distance. On the one hand, the calculation of the rotation error is complex, which will take up a lot of CPU resources; on the other hand, the rotation error is very small in a short time range, so the rotation error can be ignored in a short range. In this way, a kind of translation transformation can be carried out in a short interval, so that the navigation track of this segment is basically consistent with the real track. Therefore, formula (2) can be optimized to

$$E = \sum_{i=1}^{N} d(P_i X_i) = \sum_{i=1}^{N} \left\| x_i - (p_i + \vec{t}) \right\| \tag{3}$$

(2) The determination of set of road points to be matched and the nearest point corresponding to the track. The ICCP algorithm for gravity, geomagnetism and terrain matching, since the track points must be on the corresponding isoline, the corresponding isoline is generally used as the matching unit, that is to find the nearest point of the track through the isoline in the map. Although there is no corresponding isoline information to be used in path matching, vector road data has good geometric attributes, which can be used for lossless mathematical transformation such as rotation, scaling, perspective [28]. Therefore, this paper uses the geometric attributes of vector road data to identify the set of road points to be matched, such as the curve opening size and curve opening direction of the road. The set of road points to be matched would be searched out from the error confidence interval allowed by the integrated navigation. At the same time, the inertial navigation system can provide high precision heading angle information, so it can find the corresponding closest point set on the road along the normal direction of the heading angle, which could be taken as the estimated real set of trajectory points.

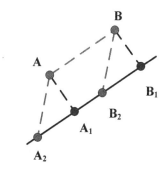

Fig. 2. Matching error of linear road

(3) The matching error of the track of the straight line segment. During the matching process, the positioning error of INS is generally divided into the positioning error along the road and the positioning error perpendicular to the road. If we only find the closest point according to the normal direction of the track point's travel direction, we can only get the positioning error perpendicular to the road, but not along the road. As shown

in Fig. 2, AB is the navigation track of the vehicle, if only the error perpendicular to the road direction is considered, the matching result would be A_1B_1, but we know that A_2B_2 would be the real track corresponding to AB. Thus the error accumulation would be increased, which will affect the accuracy of subsequent matching. Therefore, it is necessary to match the road sections with turning feature, so as to obtain the positioning errors in two directions.

Generally, the algorithm idea of this paper can be summarized as follows: searching the corresponding road features according to the geometric similarity of vehicle track features, finding the closest corresponding point of the track point, calculating the translation transformation amount, and carrying out iterative matching to get a correction track which is consistent with the corresponding road. There are three key problems in the algorithm: choosing the set of road points to be matched, finding the nearest point on the road corresponding to the trajectory, and calculating the amount of translation for iterative matching.

3 Description and Construction of Algorithm

3.1 Definition of Terms

It can be seen from Sect. 2.2 that the matching process is carried out in the road section with certain bending feature. Therefore, this paper adopts the trend set trajectory feature division algorithm based on the change of heading angle proposed by the author in reference [29], and extracts the trajectory and road features in the form of sequence point set according to the change of vehicle heading angle and road azimuth angle, which are respectively recorded as $\{P\}$, $\{Q\}$. $P = \{P_i, \quad i = 1, 2, 3, \cdots, m\}$, $Q = \{Q_i, \quad i = 1, 2, 3, \cdots, n\}$, in which m and n represent the number of track points or road points contained in the set of feature points.

During the driving, vehicles generally travel along the road, so the track features and road features have certain geometric similarity. According to the geometric properties of features, the definition is as follows:

Definition 1: the first point in the sequence point set of feature curve is called **the first node of feature**; the last point in the sequence point set is called **the last node of feature**.
Definition 2: the point with the largest curvature of the feature curve or the point with the same length from the first and last nodes of the feature is called **the feature vertex**.
Definition 3: the counterclockwise angle between the perpendicular direction of the line between the vertex of the characteristic curve and the beginning and end point of the characteristic curve and the north direction is called **the opening direction of the curve (feature)**.
Definition 4: the angle between the vertex of the feature curve and the line between the beginning and the end of the feature is called **the opening size of the curve (feature)**.

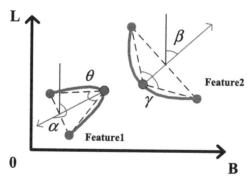

Fig. 3. Opening direction and size of curve (feature)

As shown in Fig. 3, the angles of α and β are respectively the opening directions of features 1 and 2, and the angle sizes of θ and γ are respectively the opening sizes of features 1 and 2.

3.2 Determination of Road Point Set to Be Matched

Using the characteristic geometric attributes given in Sect. 3.1, the method of curve fitting is given in reference [30] to choose the set of road points to be matched. However, for the nonlinear curve or complex multivariate curve, the fitting process needs a lot of calculation and analysis, such as rotation transformation, which have a certain extent affects to the efficiency of algorithm. Therefore, this paper uses the opening direction, size of the characteristic curve and the confidence interval of navigation error to search the set of road points to be matched.

According to the feature of vehicle trajectory, the minimum circumscribed rectangle of the region where the trajectory features are located can be determined. Considering the position error of INS, it is necessary to enlarge the extracted boundary properly. The search range of the path to be matched in the trajectory feature set {P} is determined by the inertial navigation positioning error ellipse [31]. It is assumed that the positioning error of INS follows the standard normal distribution, σ_x^2, σ_y^2 and σ_{xy} are the East variance, North variance and covariance of INS. Then the error ellipse is calculated as follows

$$a = \sigma_0 \sqrt{(\sigma_x^2 + \sigma_y^2 + \sqrt{(\sigma_x^2 - \sigma_y^2)^2 + \frac{4\sigma_{xy}^2}{2}})} \qquad (4)$$

$$b = \sigma_0 \sqrt{(\sigma_x^2 + \sigma_y^2 - \sqrt{(\sigma_x^2 - \sigma_y^2)^2 + \frac{4\sigma_{xy}^2}{2}})} \qquad (5)$$

$$\varphi = \frac{\pi}{2} - \frac{1}{2} \arctan(\frac{2\sigma_{xy}}{\sigma_x^2 - \sigma_y^2}) \qquad (6)$$

In formula (4), (5), (6), a and b are the long semi-axis and the short semi-axis of INS error ellipse respectively; φ is the angle between the long semi-axis and the north direction of the ellipse. According to the "3σ" principle, the probability that the measured point is located in the location error ellipse is 95% when $\sigma_0 = 3.03$ [32].

In order to facilitate the calculation, the formula of error ellipse circumscribed rectangle derived in reference [31] is adopted

$$\begin{cases} x_m = 2\sqrt{a^2 \sin^2 \varphi + b^2 \cos^2 \varphi} \\ y_m = 2\sqrt{a^2 \cos^2 \varphi + b^2 \sin^2 \varphi} \end{cases} \tag{7}$$

If the boundary of track feature is B_{left}, B_{right}, L_{down} and L_{top}, and there is $\{P\} = \{P|(\lambda_i, L_i)|i = 0, 1, \cdots, m\}$, (γ_i, L_i) represents the longitude and latitude coordinates of track points in track feature set P, then there is

$$B_{left} = \min(\lambda_i) - x_m \tag{8}$$

$$B_{right} = \max(\lambda_i) + x_m \tag{9}$$

$$L_{down} = \min(L_i) - y_m \tag{10}$$

$$L_{up} = \max(L_i) + y_m \tag{11}$$

In Eq. (8) (9) (10) (11), $\max|\cdot|$ and $\min|\cdot|$ represent taking the maximum value and taking the minimum value respectively.

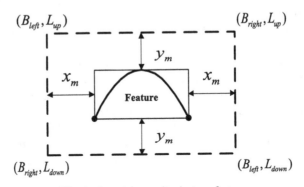

Fig. 4. Search area of trajectory feature

As shown in Fig. 4, the outside border of the dotted line is the area of the road point set to be matched searching by track feature. If a set of road feature points Q falls in this search area, it is necessary to further determine whether the opening direction and opening size of the feature curve meet the tolerance requirements. Therefore, α_P and α_Q are the opening directions of track feature and road feature are respectively, and β_P and β_Q are the opening sizes of track feature and road feature are respectively. Then there are

$$\begin{cases} \Delta\alpha = |\alpha_P - \alpha_Q| \\ \Delta\beta = |\beta_P - \beta_Q| \end{cases} \tag{12}$$

In Eq. (12), $|\cdot|$ represents the absolute value.

Because the vehicle will be affected by factors such as emergency avoidance or drivers' habits during the driving, there will be some differences between the track and the opening direction and size of the road. Take two-way eight lanes as an example, the lane width is 2.5 m–3 m, considering that the maximum driving distance of vehicles in the turning process is about 50m, so there is $\sin \Delta \alpha = 0.18$, the maximum curve opening direction limit is $\Delta \alpha = 10°$; $\Delta \beta = 10°$ is same as above.

Therefore, when and only when $\Delta \alpha \leq 10°$ and $\Delta \beta \leq 10°$, Q is the set of road feature points to be matched corresponding to the set of track feature points P. At this time, it is noted that the number of feature points set of road to be matched is N. if N = 0, the corresponding road feature to be matched is not found, this track feature would be skipped and the road feature to be matched for the next track feature could begin to search. If N = 1, we would not deal with it; if N > 1, we need to select the best condition in the following paper.

3.3 Find the Closest Point on the Road Corresponding to the Track Feature

The actual road data is not a continuous curve, but consists of many line segments. Each line segment takes two adjacent positions in the path as the end point, which is recorded as $\{r_j, r_{j+1}\}$. As shown in Fig. 5, the blue dot set is the track feature set P, and the corresponding black dot set is the road feature set Q. In order to ensure the heading consistency of the corresponding point of the track point, this paper takes the intersection of the current heading normal of the track point A and $\{r_j, r_{j+1}\}$ as its corresponding point on the road, which is recorded as A_0. If there is no intersection between the normal and Q, the track point is ignored; if there is more than one intersection, the intersection closest to Pi is taken as the effective corresponding point.

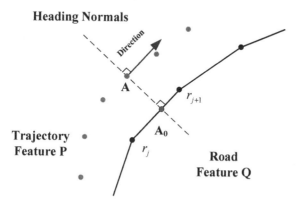

Fig. 5. Corresponding points of the trajectory points

3.4 Calculation of Translation for Iterative Matching

3.4.1 Calculation of Translation

After acquiring the corresponding point A_0 of track point in road feature, the horizontal and vertical translation from track point a to road point A_0 can be expressed as

$$t_{x_i} = \lambda_0 - \lambda_i \tag{13}$$

$$t_{y_i} = L_0 - L_i \tag{14}$$

According to the method in Sect. 3.3, (λ_i, L_i) represents the coordinates of A_i, the distance from each track point to its road corresponding point could be obtained by traversing the set P of track feature points. If no corresponding point is found for the track point, the distance from the track point to the road is recorded as 0. Therefore, the average distance in longitude and latitude from all the track points in {P} to the road feature can be expressed as:

$$\bar{t}_x = \frac{\sum\limits_{i=0}^{k} t_{x_i}}{k - m} \tag{15}$$

$$\bar{t}_y = \frac{\sum\limits_{i=0}^{k} t_{y_i}}{k - m} \tag{16}$$

In Eq. (15), (16), k is the number of track points in track feature P, and m is the number of track points in P that can not be found.

3.4.2 Iterative Matching and Setting of Its End Conditions

According to the translation amount of Eq. (15), (16), track feature P_i is translated along longitude direction and latitude direction respectively $\bar{t}_x$ and $\bar{t}_y$, and a matching correction is completed to obtain a set of updated track points P_i', namely

$$P_i' = TP_i = \vec{T} + P_i \tag{17}$$

After that, the new set P_i' is used as the starting set to repeat the steps of Sects. 3.3 and 3.4 until the convergence, until T would be stable and no obvious change occurs. At this time, the set of trajectory points TP obtained after the last convergence iteration is the final matching trajectory.

Considering the memory requirements and system efficiency of the computer, this paper takes 30 times of iterative translation as the upper limit of iteration, and if more than 30 times, the matching correction ends. In addition, when the approximation degree of the track point is basically consistent with the road point, that is the deviation of longitude or latitude is less than the allowable basic error range of navigation, it is also considered that the matching is successful. At this time

$$\sqrt{\bar{t}_x^2 + \bar{t}_y^2} < d_{\max} \tag{18}$$

$\mathbf{d_{max}}$ in Eq. (18) is the maximum error required by integrated navigation, and is set to $0.90901 \times 10^{-4\circ}$ (about 10 m) according to the error requirements of general navigation.

As shown in Fig. 6, after the translation matching correction meets the iteration conditions, the track points are basically consistent with the road points, meeting the requirements of navigation matching.

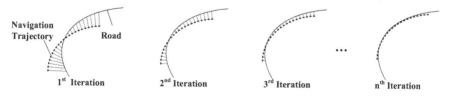

Fig. 6. Process of iterative matching

3.4.3   Construction of Objective Function and Calculation of Matching Variance

In general, the objective distance function constructed in this paper is as follows:

$$E = \sqrt{\overline{t_x'}^2 + \overline{t_y'}^2} \times k$$
$$= \sqrt{(\frac{\sum\limits_{i=0}^{k} t_{x_i}}{k-m})^2 + (\frac{\sum\limits_{i=0}^{k} t_{y_i}}{k-m})^2} \times k \qquad (19)$$

The target distance function E represents the sum of the distance difference between the track point and the estimated real track point. After iterative calculation, the E value gradually decreases and reaches convergence, indicating that the matching point is more and more close to the real track point. Therefore, if E decreases gradually and converges at the end of iterative matching, the matching is considered successful, otherwise, the matching fails and the algorithm ends.

If the target distance function converges, the matching results need to be evaluated. The matching variance mainly describes the matching degree between the navigation track and the road after the path matching. Its size directly reflects the advantages and disadvantages of the matching results, and at the same time, it can filter out the optimal matching track results.

The general calculation formula of matching variance δ^2 is:

$$\delta^2 = \frac{1}{n} \sum_{i=1}^{n} ((x_i - x_i')^2 + (y_i - y_i')^2) \qquad (20)$$

Where (x_i, y_i) is the coordinate of the track point after the last translation in the iteration process, and (x_i', y_i') is the corresponding point of the track point on the road. As there may be some track points without corresponding road points, the specific calculation method of matching variance is as follows:

(1) In Sect. 3.3, if the track points have road feature corresponding points, the distance square between the track points and their corresponding points after iterative translation is calculated and recorded as DIS^2; otherwise, the number of track points without corresponding points is recorded as m;

(2) Find the maximum value of DIS^2; and record it as set D^2_{max};

(3) When the iterative translation matching is completed, the sum of the square of the distance from all the track points to their corresponding points could be calculated. If the square of the distance from the track points where the matching points are not found is uniformly specified as D^2_{max}, then

$$DIS^2_i = \begin{cases} (x_i - x'_i)^2 + (y_i - y'_i)^2, & \textit{Corresponding points} \\ D^2_{max}, & \textit{No corresponding points} \end{cases} \qquad (21)$$

Then the calculation formula of matching variance can be expressed as follows:

$$\delta^2 = \frac{1}{n}(\sum_{i=1}^{n-m} ((x_i - x'_i)^2 + (y_i - y'_i)^2) + m \times D^2_{max}) \qquad (22)$$

Where n is the total number of track points in the set of track feature points, and m is the number of track points for which no feature points are found. The smaller the value is, the closer the matched track points are to the real track points, and the better the matching result is.

Therefore, in Sect. 3.2, if the number of road feature point sets to be matched N > 1, after the iterative matching, the minimum correction result of the corresponding road feature point set is taken as the final matching result.

4 Implementation of Algorithm

According to the analysis of the above algorithm, the key technical process of this algorithm is shown in Fig. 7.

The specific steps of the algorithm are as follows:

Step 1. The algorithm initializes, reads the vector road point data, receives the navigation track point data, and extracts the feature sets of track and road data respectively.

Step 2. Determine the set of road points to be matched according to the opening direction, size and error confidence interval of track characteristics. If the number of road points to be matched n > 0, turn to **Step3**; otherwise, turn to step1.

Step 3. Find the nearest point on the road according to the normal direction of the heading angle of the current track point.

Step 4. Calculate the average translation amount of the whole set of track points according to the corresponding nearest point.

Step 5. Translate the set of trajectory points as a whole according to the average amount of translation, get a new set of trajectory points, and judge whether the iteration end condition is met. If the conditions are met, turn to step6; otherwise, turn to **Step3**.

Step 6. Calculate the matching variance to judge and output the optimal set of matching trajectory points.

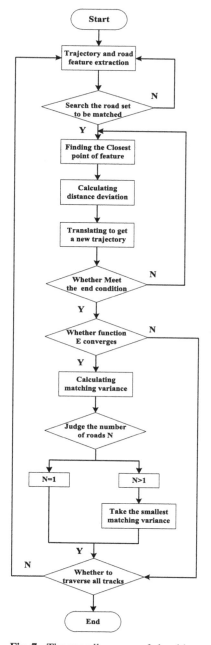

Fig. 7. The overall process of algorithm

5   Experiment and Result Analysis

5.1   Algorithm Feasibility Analysis

In order to verify the feasibility and matching effect of the algorithm, this paper carries out a field sports car experiment on the road section of about 30 km in Jinyuan Bridge area of the Fifth Ring Road in Beijing. The collected data is the actual trajectory data of inertial navigation and the high-precision vector road data. Among them, the average speed is 60 km/h, the gyro constant drift is 0.05 °/h, the random drift is 0.01 °/h, the average value of the initial zero deviation of the accelerometer is 10^{-4} g, the random zero deviation is 0.5×10^{-4} g, and the initial value of the East, North and azimuth misalignment angle is $1°$ [29]. Because the vector road data is the result of difference, the influence caused by low sampling frequency is not considered temporarily. Based on the Visual C++ 6.0 platform, this paper also uses the algorithm to carry out the simulation experiment of matching and modifying the track points. The simulation results are shown in Fig. 8, 9, 10, 11, 12.

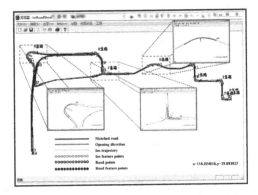

Fig. 8. Simulation results of ICCP road matching

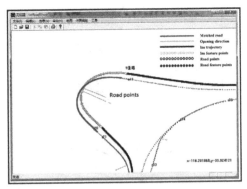

Fig. 9. Partial enlarged in ICCP road matching 1 (when diverged angle less than $90°$)

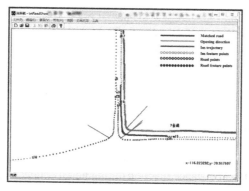

Fig. 10. Partial enlarged in ICCP road matching 2 (when diverged angle is 90°)

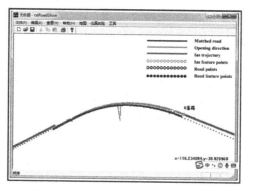

Fig. 11. Partial enlarged in ICCP road matching 1 (when diverged angle more than 90°)

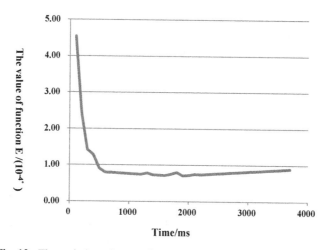

Fig. 12. The variation of target distance function with matching process

Figure 9, Fig. 10 and Fig. 11 show the actual track, road and matching track after algorithm iterative translation of the vehicle. The blue track in the figure is the inertial navigation track. It can be seen that the red matching track can better locate and track the black road. It can be seen from the partial enlarged figure in Fig. 10 that after several iterations of translation, the matching track and the shape of the road are basically consistent. As the track points move forward, the matching points are more and more close to the road, that is to say, the algorithm has reached convergence. Figure 11 shows the matching effect of a road intersection. Through the matching results, it can be seen that the algorithm in this paper can not only complete the matching at the turning of the road, but also have a good matching effect at the complex road intersection. Figure 12 shows the change of target distance function during iterative matching. As time goes on, its function value tends to be stable and converges. ICCP algorithm is initially applied to the image registration. Through the above simulation experiments, it can be concluded that the algorithm in this paper is completely feasible for road matching aided INS.

5.2 Algorithm Iteration Threshold Analysis

In Sect. 3.4.2, the upper limit of the number of iterations is set to 30 in the process of matching calculation. In order to verify the rationality of the iterative conditions, five navigation tracks in Jinyuan Bridge area of the Fifth Ring Road in Beijing are simulated by path matching, and different iterations are set respectively. In order to facilitate the comparison and analysis of the results, the matching standard deviation is used as the evaluation standard. In Table 1, the matching standard deviation, iteration time and the total number of track points corresponding to different iterations are calculated.

Table 1. Matching results and time of algorithm with different iteration times

Number of iterations	Evaluation standard	Track No					
		1	2	3	4	5	Average
10	Matching standard deviation $\delta(\times 10^{-4})$	1.19444	1.35370	1.32716	1.35697	1.28735	1.30392
	Iteration time/(s)	2.7	0.7	1.1	0.5	2.4	1.5
30	Matching standard deviation $\delta(\times 10^{-4})$	0.88416	0.90405	0.90574	0.90631	0.89827	0.89971
	Iteration time/(s)	6.5	1.7	2.5	1.1	4.7	3.3
50	Matching standard deviation $\delta(\times 10^{-4})$	0.80379	0.82187	0.82340	0.82392	0.81661	0.81792
	Iteration time/(s)	15.6	6.4	8.0	5.2	12.7	9.6
Number of track points		3660	862	1148	521	2560	1750

According to Table 1, with the increase of the number of iterations, the matching standard deviation is shrinking, and the matching results are more and more ideal. However, with the increase of the calculation amount, the iteration time is also increasing. When the number of iterations is 10, although the iteration time is short, the matching standard deviation is large, which does not meet the system requirements of $0.90901 \times$

10^{-4} in Sect. 3.4.2; when the number of iterations is 30, the matching standard deviation meets the requirements of the system, and the iteration time increases stably. When the number of times is 50, the matching standard deviation is further reduced and the matching result is ideal. However, compared with the impact of the large increase of iteration time on the overall efficiency of the algorithm, the improvement of the 10^{-6} magnitude of the matching standard deviation is not significant. In conclusion, it is reasonable to select 30 iterations.

5.3 Comparison and Analysis with Traditional Algorithm

Finally, this paper uses the direct projection method, the weight matching method and the algorithm in this paper to carry out the path and INS data matching simulation experiments on five tracks of different sections in Jinyuan Bridge area of the Fifth Ring Road in Beijing, and calculates the matching standard deviation, the matching accuracy and the time consumed by the three methods respectively. The definition of the matching accuracy is as follows:

Matching accuracy = number of correctly matched track feature points / total number of navigation track feature points.

The experimental statistical results are shown in Table 2.

Table 2. Matching results of 3 algorithms

Algorithm	Evaluation standard	Track no.					
		1	2	3	4	5	Average
Our algorithm	Accuracy rate/(%)	91.9	90.1	87.6	93.1	89.8	90.5
	Time/(s)	6.5	1.7	2.5	1.1	4.7	3.3
	Matching standard deviation ($\times 10^{-4}$)	0.88416	0.90405	0.90574	0.90631	0.89827	0.89971
Projection algorithm	Accuracy rate/(%)	76.4	82.7	78.9	80.1	77.9	79.4
	Time/(s)	5.8	1.5	2.3	1.0	4.2	3.0
	Matching standard deviation ($\times 10^{-4}$)	1.19362	1.22047	1.22275	1.22352	1.21267	1.21461
Weights algorithm	Accuracy rate/(%)	89.3	84.1	86.2	82.5	86.5	85.7
	Time/(s)	11.7	3.1	4.5	2.0	8.5	6.0
	Matching standard deviation ($\times 10^{-4}$)	0.89301	0.91309	0.91480	0.91537	0.90725	0.90870
Number of track points		3660	862	1148	521	2560	1750

According to the analysis in Table 2, the average matching accuracy of the algorithm in this paper can reach 90.5%, which also meets the requirements of positioning error. Compared with the traditional vertical projection algorithm and weight algorithm, the algorithm improves and effectively reduces the probability of false matching. In terms of matching time, due to the limitation of iteration times, the matching time is slightly longer than that of the direct projection algorithm. But compared with the weight algorithm, the matching efficiency has been improved. In conclusion, our algorithm has practical value in path matching.

In addition, this paper also compares the five indicators of the three algorithms. Due to the limited space, the detailed statistical information of the matching results will not be listed, and the performance comparison relationship is shown in the figure in Fig. 13.

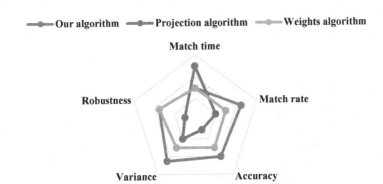

Fig. 13. Performance comparison of three algorithms

6   Conclusion

In this paper, we propose a vector road data aided vehicle inertial navigation algorithm based on ICCP, which takes high-precision vector road data as a new auxiliary information, and corrects the inertial navigation error by this algorithm. The algorithm can be applied to the aid inertial navigation system without GPS signal, which can effectively restrain the divergence of the trajectory error of the inertial navigation system, and realize the auxiliary positioning without satellite navigation.

(1) The matching trajectory obtained by iterative translation is basically consistent with the real road, and the location of the vehicle is effectively tracked, which shows that ICCP algorithm is feasible in road matching aided INS navigation and positioning.
(2) The setting of the number of matching iterations has a direct impact on the accuracy and efficiency of matching. Experiments show that when the number of iterations is 30, the matching variance and iteration time of the algorithm in this paper reaches the optimal combination;
(3) The performance of the algorithm is verified by simulation. The experimental results show that the accuracy and efficiency of this algorithm are better than the traditional direct projection algorithm and the weight algorithm, and the matching results are ideal, which can meet the requirements of integrated navigation matching.

References

1. Li, X.: Research on technology and method of vector road data aided inertial navigation. Acta Geodaet. et Cartographica Sin. **47**(5), 692 (2018)

2. Li, B., Yao, D.: Low-cost MEMS IMU navigation positioning method for land vehicle. J. Chin. Inertial Technol. **22**(6), 20–24 (2014)
3. Yan, Y., Li, F.: Research of rough map matching algorithm based on GPS/DR integrated navigation. Comput. Eng. Appl. **51**(1), 261–265 (2015)
4. Guo, L., Li, H., Zhang, Z., et al.: Geometry matching method for transportation road network data based on projection. Geom. Inf. Sci. Wuhan Univ. **38**(9), 1113–1117 (2013)
5. Liu, F., Balazadegan, Y., Gao, Y.: Tight integration of INS/stereo VO/digital map for land vehicle navigation. Photogram. Eng. Remote Sens. **84**(1), 15–23 (2018)
6. Hao, L., Deng, Z., Fu, M., et al.: A map matching algorithm for inertial navigation systems based on the adaptive projection method. In: Seventh International Symposium on Computational Intelligence & Design (2015)
7. Zhao, D., Liu, X., Guo, L.: Real time map matching algorithm of floating car in support of spatial grid index. J. Comput.-Aided Des. Comput. Graph. **26**(9), 1550–1556 (2014)
8. Su, H.B., Chen, Y., Liu, Q.: A map matching algorithm based on weight. J. Noah China Inst. Water Conserv. Hydroelectr. Power **29**(1), 81–83 (2008)
9. Newson, P., Krumm, J.: Hidden Markov map matching through noise and sparseness. In: Proceedings of the 17th ACM SIGSPATIAL International Conference on Advances in Geographic Information Systems, pp. 336–343 (2009)
10. Yuan, J., Zheng, Y., Zhang, C., et al.: An interactive-voting based map matching algorithm. In: MDM, pp. 43–52 (2010)
11. Li, X., Hua, Y., Zhang, H., et al.: Positioning correction algorithm of vector road aided inertial navigation based on the rough/fine matching method. Acta Geodaet. et Cartographica Sin. **46**(8), 1034–1046 (2017)
12. Pinto, A.M., Moreira, A.P., Costa, P.G.: A localization method based on map-matching and particle swarm optimization. J. Intell. Rob. Syst. **77**(2), 313–326 (2015). https://doi.org/10.1007/s10846-013-0009-2
13. Imprialou, M.M., Quddus, M., et al.: High accuracy crash mapping using fuzzy logic. Transp. Res. Part C **42**(2), 107–120 (2014)
14. Che, M., Wang, Y., Zhang, C., et al.: An enhanced hidden markov map matching model for floating car data. Sensors **18**(6), 1758 (2018)
15. Besl, P.J., McKay, N.D.: A method for registration of 3-D shapes. IEEE Trans. Pattern Anal. Mach. Intell. **14**(2), 239–256 (1992)
16. Chen, Y., Medioni, G.: Object modeling by registration of multiple range images. Image Vis. Comput. **10**, 145–155 (1992)
17. Bergevin, R., Soucy, M., Gagnon, H., et al.: Towards a general multi-view registration technique. IEEE Trans. Pattern Anal. Mach. Intell. **18**(5), 540–547 (1996)
18. Behzad, K.P., Behrooz, K.P.: Vehicle localization on gravity maps. In: Proceedings of SPIE-The International Society for Optical Engineering, Orlando, Florida, vol. 3693, pp. 182–191 (1999)
19. Liu, F.M., Sun, F., Cheng, Y.: Gravity localization based on ICCP algorithm and its generalization. J. Chin. Inertial Technol. **12**(5), 36–39 (2004)
20. Wang, Z., Bian, S.: ICCP algorithm for gravity aided inertial navigation. Acta Geodaet. et Cartographica Sin. **37**(2), 147–151 (2008)
21. Yan, L., Cui, C., Xie, H.: The application of ICCP algorithm to gravity matching. Remote Sens. Inf. (1), 16–19 (2009)
22. Xiao, S., Bian, S., Huang, X.: Research on geomagnetism aided inertial navigation by using ICCP algorithm. Ship Electron. Eng. **31**(5), 83–86 (2011)
23. Song, L.: Research on the multi-level geomagnetic match algorithm aided inertial navigation based on entropy/ICCP algorithm. J. North China Inst. Sci. Technol. **11**(10), 70–74 (2014)

24. Liu, R., Zhang, J., Li, X.: The research on the improvement of ICCP algorithm for the route-matching of initial navigation vehicles in road network. J. Geomat. Sci. Technol. **33**(1), 82–87 (2016)

25. Li, X.: Research and implementation on aided inertial navigation positioning algorithm based on road networks feature. Information Engineering University. Department of Geography Information Engineering, Zhengzhou (2013)

26. Sobreira, H., Costa, C.M., Sousa, I., et al.: Map-matching algorithms for robot self-localization: a comparison between perfect match, iterative closest point and normal distributions transform. J. Intell. Robot. Syst. **93**, 533–546 (2019). https://doi.org/10.1007/s10846-017-0765-5

27. Wang, X., Su, M., Liu, P., et al.: Application of improved ICCP algorithm in gravity matching aided navigation. Sci. Surv. Mapp. **38**(1), 36–39 (2013)

28. Sun, Y., Huang, B., Wang, L., et al.: Vector map matching navigation method with anti-scale transformation. J. Chin. Inertial Technol. **21**(1), 89–91 (2013)

29. Li, X., Zhang, J., Yang, B., et al.: An extraction algorithm of track features based on trend set of heading angle variable. J. GeoInf. Sci. **17**(10), 1172–1178 (2015)

30. Li, X., Zhang, J., Ge, S., et al.: An assistant navigation positioning algorithm based on the road curve feature and inertial navigation data. J. Geomat. Sci. Technol. **30**(2), 210–213 (2013)

31. Shen, J., Zhang, J., Zha, F.: Improved ICCP algorithm for underwater terrain matching method. J. Huazhong Univ. Sci. Technol. (Nat. Sci. Ed.) **40**(10), 63–67 (2012)

32. Sheng, Z., Xie, S., Pan, C.: Probability Theory and Mathematical Statistics, 4th edn., pp. 161–163. Higher Education Press, Beijing (2008)

Traffic Flow Forecasting Using a Spatial-Temporal Attention Graph Convolutional Network Predictor

Shan Jiang[1,2], Meiling Zhu[2(✉)], and Jin Li[2]

[1] University of Chinese Academy of Sciences, Beijing, China
jiangshan2018@iscas.ac.cn
[2] Institute of Software, Chinese Academy of Sciences, Beijing, China
{meiling,lijin}@iscas.ac.cn

Abstract. Traffic flow forecasting is a hotspot in the field of the smart city. It is a highly nonlinear, complex, and dynamic problem affected by many factors. The traditional methods cannot well model the dynamic spatial-temporal correlations of long-range time series data in the traffic network, which reduces the forecast accuracy. In this paper, we proposed a novel attention-based graph neural network predictor to forecast traffic flow. A more flexible and efficient convolution operation is defined in our predictor based on graph wavelet transform. The predictor can capture both spatial and temporal relationships based on a graph wavelet neural network. More specifically, the network adopts the spatial-temporal attention mechanism to capture the dynamic spatial-temporal correlations, and the dilated 1D convolution component is stacked to handle long sequences. We made experiments on two real-world benchmark datasets to verify the accuracy of the proposed network.

Keywords: Traffic flow forecasting · Graph wavelet · Spatial-temporal graph convolution network

1 Introduction

Recently, intelligent transportation systems are developed to build smart cities. Traffic flow forecasting is a key ability of intelligent transportation systems, which is widely applied to relieving congestion. Based on accurate traffic flow forecasting, intelligent transportation systems can provide vehicle guidance and congestion relief to the traffic management department. Traffic flow forecasting is a typical problem of temporal and spatial data prediction [1–3].

Quick and accurate traffic flow forecasting is not an easy task due to its highly nonlinear, complex, and spatial-temporal dynamic features [4]. Figure 1 is an example for describing the spatial-temporal dynamics of road graphs. Figure 1(a) shows a typical

This work is supported by the National Natural Science Foundation of China under Grant Nos. (61703013, 91646201).

X. Meng et al. (Eds.): SpatialDI 2020, LNCS 12567, pp. 107–121, 2021.
https://doi.org/10.1007/978-3-030-69873-7_8

crossroad network. As Fig. 1(a) shows, ①–⑧ are lanes, while dotted arrows describe the driving rules among the eight lanes. For example, vehicles can move from lane ② ⑥ or ⑧ to lane ⑤. Figure 1(b) is an undirected road graph extracted from (a). If an event occurred on lane ⑤, the traffic flow from lane ② ⑥ or ⑧ to lane ⑤ will be blocking. Under these circumstances, we can extract another graph Fig. 1(c). What's worse is lane ⑦ is likely to turn to block. The corresponding graph is shown in Fig. 1(d). Road network have different structures in different time ranges and space range. An intuitive example is listed in Fig. 1. The graph structure is sparse during the rush hour in the city center owing to traffic jams, while it becomes dense in the whole city in the midnight. Therefore, how to capture the dynamic spatial-temporal dependencies among nodes and edges becomes a critical task in traffic flow forecasting.

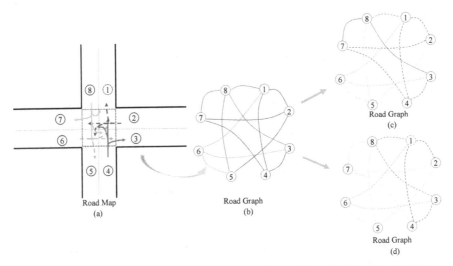

Fig. 1. A typical crossroad network (a) from which a road network graph (b) is extracted. When an event has occurred on road ⑤, e.g., the traffic jam, the car accident, the parade, etc., the connection is forced offline, and a new road graph (c) is formed. There is a risk of traffic tensions worsening when the road ⑤ is offline, e.g., another jam happened on road ⑦ (c).

In recent years, many devices have been deployed on highways to collect traffic information. Each device is placed in a fixed geographic location and continuously generates time-series data. It lays a data foundation for traffic flow forecasting. To address the spatial-temporal dynamics in the traffic flow forecasting issue, this paper presented a novel attention-based graph learning network predictor, named spatial-temporal attention-based graph wavelet neural network (STAGWNN). Our work was based on the efforts of [4], their model mainly consists of three independent components to respectively model three temporal properties of traffic flows, i.e., recent, daily-periodic, and weekly-periodic dependencies. Their model is complex and is not conducive to expanding to other modules. Different from [4], in this paper, we developed an attention-based graph learning network predictor by leveraging heat kernel diffusion wavelet patterns [5, 6] and spatial-temporal attention mechanism. The network stacked submodules to capture the temporal dependencies of recent periodicity in traffic data straightforwardly. We defined a more flexible and efficient convolution operation on the graph in our effort. Moreover, the proposed method is highly efficient by avoiding eigenvalue decomposition of the Laplacian matrix, and the graph wavelet transform matrix can be easily obtained via a fast algorithm [7] without requiring a costly matrix eigenvalue decomposition. Furthermore, the proposed model is localized in the node domain, representing the information diffusion centered at each node [8]. Finally, two real-world traffic datasets were used to evaluate our traffic flow forecasting predictor proposed in this work. The experimental results showed that our predictor outperforms previous baselines in the traffic flow forecasting task. The main contributions of this work are summarized as follows:

- We utilized graph wavelet diffusion kernels to achieve localized correlations in the spatial domain of the traffic network to predict traffic flow.
- We developed a novel attention-based graph learning network that can capture both spatial and temporal relationships via a graph wavelet neural network.
- We evaluated our proposed model on two real-world benchmark datasets of road network and achieved the best forecasting performances compared with the existing baselines.

The remainder of this paper is organized as follows. Section 2 gives a brief overview of the related works of traffic forecasting. Section 3 reports more details of the traffic flow forecasting predictor proposed in this paper. Section 4 presents the experimental evaluation results. The last section draws conclusions.

2 Related Works

After years of uninterrupted efforts, many findings have been achieved in traffic forecasting research. These methods are successful based on some specific data assumptions, but the traffic flow data is too complex to satisfy these assumptions. The traditional studies of traffic forecasting were widely based on limited survey data and focused on the empirical analysis of key factors. For instance, [9] and [10] mainly used autoregressive integrated moving average (ARIMA) and support vector regression (SVR) [11] to predict incoming traffic flow without considering the entire spatial-temporal features in the traffic network and perform not very well. Over the last ten years, some deep learning approaches for traffic flow forecasting were developed such as long short-term memory network (LSTM) [12], gated recurrent unit network (GRU) [13], etc., and showed good performance. These methods have a complex network structure that can fit nonlinear dependencies in traffic flow sequences, but it cannot satisfy the long-term forecasting task. [14] was the first to introduce graph convolutional recurrent network to capture both spatial structures and dynamic variation from structured time series data. The key drawback of this project was the lack of the ability to model the dynamic spatial-temporal correlations of traffic data. A reasonable approach to tackle this issue is to determine the optimal combinations of recurrent networks and graph convolution under specific settings. To this end, [15] introduced the gated recurrent units to graph convolution for long-term traffic flow prediction. In contrast to the aforementioned works mentioned above, [16] developed a novel graph deep learning model known as STGCN, which integrate graph convolution with gated temporal convolution through spatial-temporal convolutional blocks. But [16] did not consider the dynamic spatial-temporal correlations of traffic data. Inspired by the attention mechanism, [4] employed graph convolution and the attention mechanism to model the spatial-temporal dependencies of traffic data. The attention-based method has acquired some initial success and offers a flexible framework to generalize CNN to graph structure, but it is still elusive to determine the appropriate neighborhood of center vertex. To tackle the above challenges, [5] presented a graph wavelet neural network to implement efficient convolution on graph data using a heat diffusion wavelet obtaining the correlative neighbors of the core node. However, the graph wavelet network cannot handle the spatial-temporal sequences data. To address the aforementioned issue, we build up our network based on graph diffusion wavelet convolution combined with the spatial-temporal attention mechanisms. Our model consists of two core components, the graph convolution based on graph wavelet transform (GCWN), and the temporal causal convolution (TCCN). To further improve the forecast accuracy of this architecture, the proposed network has embedded an attention mechanism model to predict traffic flow accurately. Furthermore, the residual and skip connection strategies can integrate into the developed network easily.

3   Spatial-Temporal Attention Graph Convolutional Network Predictor for Traffic Flow Forecasting

In this section, we present a novel traffic flow forecasting method, coined spatial-temporal attention graph convolutional network (STAGWNN). Our traffic flow predictor is based on the graph structure of the traffic road network and integrates a powerful spatial-temporal feature extractor. To improve the forecast accuracy of this predictor, the attention mechanism is embedded into the proposed predictor to forecast traffic flow accurately.

3.1   Definitions

Definition of Traffic Flow Forecasting Problem. We present the formal definition of the traffic network and traffic flow forecasting as following.

Definition 1 (Traffic Graph). A road graph is modeled as a graph $G = (V, E, A)$ in this paper, where V refers to the set of $|V| = N$ nodes; E refers to the set of edges, indicating the connectivity between the nodes; $A \in R^{N \times N}$ is the adjacency matrix with $a_{ij} = a_{ji}$ which defines the connections between node i and node j. $L = D - A$ is the Laplacian matrix of G, D is a diagonal degree matrix with $D_{ii} = \sum_j A_{ij}$, the normalized L is defined as $\hat{L} = I_N - D^{-1/2} A D^{-1/2}$, where I_N denotes the identity matrix.

Definition 2 (Traffic Flow Forecasting Problem). Each node on the traffic graph G detects F observations (e.g. speed, traffic flow, occupy and so on) with the same frequency, that is, node i denotes a feature vector $x_t^i \in R^F$ at time step t. Given a set of historical observations $(x_t, x_{t-1}, \ldots, x_{t-M+1})$ and traffic network G, the prediction of traffic flow in the next H time steps is defined as.

$$\hat{x}_{t+H}, \ldots, \hat{x}_{t+2}, \hat{x}_{t+1} = f(x_t, x_{t-1}, \ldots, x_{t-M+1}), \tag{1}$$

where $x_t \in R^{N \times F}$ is an observation feature matrix at time step t, each entry of which records historical measurements for the road segment.

Graph Convolution. The majority of spectral graph convolution is defined based on the graph Fourier transform as below [17]

$$x *_G y = U((U^T y) \odot (U^T x)) = U g_\theta U^T x, \tag{2}$$

where $*_G$ represents the graph wavelet convolution operator, y denotes the filter kernel, $U^T x$ denotes the graph Fourier transform, $\odot$ denotes the Hadamard product operator, g_θ is a diagonal matrix of graph convolution filter.

However, there are several limitations in Eq. (2). Firstly, it is not easy to obtain the Fourier basis U which is of high computational cost with $O(n^3)$. Secondly, the basis U is a dense matrix so that the efficiency of graph Fourier transform is massive inefficiencies. Lastly, the graph Fourier transform is not localized in the node domain. Despite these limitations, [18] find a way to get the convolution kernel g_θ depending on the K-order truncated Chebyshev polynomial expansion

$$g_\theta = \sum_{k=0}^{K-1} \theta_k T_k(\hat{\Lambda}), \tag{3}$$

where $\hat{\Lambda} = \text{diag}(\lambda_1, \ldots, \lambda_n)/\lambda_{max} - I_N$, I_N denotes the identity matrix, λ_{max} is the max eigenvalue $\hat{\Lambda} = \text{diag}(\lambda_1, \ldots, \lambda_n)/\lambda_{max} - I_N$, I_N of the Laplacian matrix L, T_k denotes the K-order Chebyshev polynomial expansion which can be obtained by the stable recurrence relation $T_k(x) = 2x T_{k-1}(x) - T_{k-2}(x)$ with $T_0 = 1$ and $T_1 = x$. However, the expandability of the parametrized K-order filter g_θ is not friendly, such as the small K implies the small receptive field and the larger K implies the larger receptive field. But the large K may not reserve the locality in the node domain.

Graph Wavelets Convolution. Different from the parametrized K-order filter g_θ in Eq. (3), we introduced a novel spectral graph convolution via replacing graph Fourier transform with graph wavelet transform to capture the spatial correlations among nodes. Graph wavelet transform employed a set of wavelets as basis, which was defined as $\psi_s = (\psi_{s1}, \psi_{s2}, \ldots, \psi_{sN})$, where the wavelet ψ_{si} corresponds to a signal on the graph diffused away from node i and the parameter s denotes the degree of scaling. A wavelet associated with a node is local in the graph G: it is centered around this node and spreads on its neighborhood so that the larger the scale s is, the larger is the spanned neighborhood. Taking advantage of this local information encoded in wavelets, the graph convolution based on graph wavelets at scale s is defined as

$$x *_G y = \psi_s \left(\left(\psi_s^{-1} y \right) \odot \left(\psi_s^{-1} x \right) \right) = \psi_s g_\theta \psi_s^{-1} x, \tag{4}$$

where g_θ is the parametrized graph convolution filter, as the multiplication of eigenvectors U with a kernel G_s,

$$\psi_s = UG_sU^T, \tag{5}$$

where $G_s = diag(g(s\lambda_i)|_{i=1:N}) \in R^{N \times N}$, $g(s\lambda_i)|_{i=1:N} = e^{s\lambda_i}$ denotes a heat diffusion kernel function [17]. The ψ_s^{-1} can be simply replaced by $G_s^{-1} = diag(g(-s\lambda_i)|_{i=1:N}) \in R^{N \times N}$ correspondingly. For just one wavelet, this reads equivalently as $\psi_{si} = UG_{si}U^T\delta_i$, δ_a is the impulse on a single vertex i. We used a polynomial approximation algorithm proposed by [6] for bypassing the diagonalization of the Laplacian to approximate the wavelet [7], as below.

$$\psi_s^{-1}f = \frac{1}{2}c_0f + \sum_{k=1}^{K} c_k T_k^*(L)f, \tag{6}$$

where $f \in R^N$ denotes the graph signal, $c_k = \frac{\pi}{2}\int_0^\pi cos(k\beta)g(s(\frac{1}{2}\lambda_{max}(cos(\beta)+1)))d\beta$, $T_k^*(L)f = 2(\frac{2}{\lambda_{max}}L - I_N)T_{k-1}^*(L)f - T_{k-2}^*(L)f$, $T_0^* = I_N$, $T_1^* = \frac{2}{\lambda_{max}}L - I_N$, I_N denotes the identity matrix, λ_{max} is the max eigenvalue of the Laplacian matrix L, T_k^* denotes the K-order Chebyshev polynomial expansion. The computational complexity of approximation is linear with the number of edges in the traffic network graph.

In this paper, we used a more generalized and parametrized filter kernel g_θ which can be defined as any diagonal matrix in Eq. (4). ψ_s and ψ_s^{-1} can obtain by the polynomial approximate wavelets with the heat diffusion kernel $g(s\lambda_i)|_{i=1:N} = e^{s\lambda_i}$ and $g(-s\lambda_i)|_{i=1:N} = e^{-s\lambda_i}$ respectively. A sample synthetic graph is showed in Fig. 2, we considered a graph describing a synthetic 2d road network that contains 25 nodes and 40 edges as an example. In the synthetic network graph, each node denotes a road segment (or a detector); each edge represents a link between two nodes. Meanwhile, we showed two wavelet bases at different scales on the synthetic graph. Compared with the small scale (Fig. 2 (left)), a large scale (Fig. 2 (right)) indicates that a large receptive field (the dotted annulus in Fig. 2) the center node has. To explore the spatial dependence in the road network, we utilize the attribute that the signal diffuses away from the center node to build the convolution (Eq. (4)) in this article.

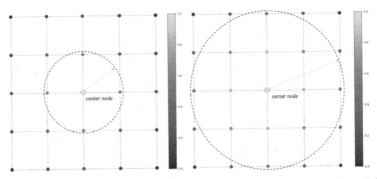

Fig. 2. Wavelets on a synthetic 2d road network (25 nodes and 40 edges) graph at (left) small scale and (right) large scale.

Temporal Convolution on Road Graphs. We adopt the 1-D dilated causal convolution [19] as our temporal convolution operation to extract the node's temporal trends in the road network graph G_t. In the dilated causal convolution networks, as the layer increased, the receptive field will become large, while preserving the input resolution as well as computational efficiency. The causal method implies that there is no leakage of information from the future into the past by padding zeros to the input sequences so that the inferences made at the current time step only contain the historical information. In this paper, we use the dilated causal convolution to capture extremely long-term temporal contextual information.

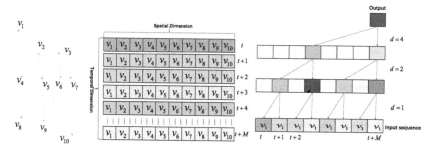

Fig. 3. Given a road network (left) which is a subgraph of the Minnesota transportation network, the dilated causal convolution is adopted to capture the node's temporal trends in the temporal dimension. An example of 1-D dilated causal convolution with the dilation rate $d = 1, 2, 4$, and the filter kernel size 2 (right) on the vertex v_1.

The 1-D dilated causal convolution slides over the input sequences by skipping values with a certain step. Given an input sequence $x \in R^M$, the dilated causal convolution operation is defined as

$$x *_D y(\text{t}) = \sum\nolimits_{k=0}^{K-1} y(\text{k})x(t - kd), \tag{7}$$

where $*_D$ denotes the dilated convolution operator, d is the dilation rate, $y \in R^K$ is the filter, t denotes the time step and K denotes the kernel size. As a special case, the dilated convolution with dilation rate $d = 1$ is equivalent to the regular convolution, as illustrated by Fig. 3. The dilated causal convolution networks can capture the long sequences with fewer layers, which can save the computation resources.

3.2 Spatial-Temporal Attention-Based Graph Wavelet Network Predictor

In this subsection, we first give the mathematical definitions of spatial attention and temporal attention. Both of them worked together to capture the dynamic spatial-temporal dependences on the traffic network graph. Lastly, we depicted the framework of our proposed predictor, known as the spatial-temporal attention-based graph wavelet network (STAGWNN) predictor for traffic flow forecasting.

Capturing the Spatial-temporal Dynamic Dependences Based on Attention. The graph convolution based on graph wavelet can extract the spatial neighbor dependencies of a center vertex and 1-D dilated causal convolution can extract the node's temporal dimension patterns, but these convolutions cannot capture the dynamic relationship of traffic flow data both the spatial and temporal dimensions. In this paper, we introduce the attention mechanism [4] to capture the dynamic spatial-temporal dependencies on the traffic network G_t both spatial dimension and temporal dimension. Mathematically, the formulation of spatial attention can be written as follows.

$$A_s = V_s \cdot h\left((X_{T_m^l}^l W_1) \cdot W_2 \cdot \left(\left(\mathbf{X}_{T_m^l}^l\right)^\mathrm{T} W_3\right) + b_s\right) = f_{A_s}\left(\mathbf{X}_{T_m^l}^l\right), \tag{8}$$

$$\hat{A}_s(i,j)|_{j=1}^N = exp(A_s(i,j))/\sum_{j=1}^N exp(A_s(i,j)), \tag{9}$$

where $A_s \in R^{N \times N}$ is the attention matrix, $\hat{A}_s$ is the standardized (or softmax) attention matrix, the element $\hat{A}_s(i,j)$ represents the strength of the connection between node i and node j, $X_{T_m^l}^l = \{X_{t-T_m-1}, \ldots, X_{t-1}, X_t\}_{l\,layer} \in R^{N \times F^l \times T_m^l}$ is the input of the l-th spatial-temporal attention layer, F is the number of channels or signals feature dimensions of the input data in l-th layer. T_m^l denotes a period of historical time series in temporal dimension at l-th layer.h denotes an activation function, e.g., a rectified linear unit (ReLU). The parameters $W_1 \in R^{T_m^l}$, $W_2 \in R^{F^l \times T_m^l}$, $W_3 \in R^{F^l}$, $V_s \in R^{N \times N}$, and $b_s \in R^{N \times N}$ are needed to be learned.

We also introduced the temporal attention mechanism to capture the dynamic correlations in the temporal dimension, as below.

$$A_t = V_t \cdot h\left(\left((X_{T_m^l}^l)^T H_1\right) \cdot H_2 \cdot \left(H_3 X_{T_m^l}^l\right) + b_t\right) = f_{A_t}\left(X_{T_m^l}^l\right), \tag{10}$$

$$\hat{A}_t(i,j)|_{j=1}^{T_m^l} = exp(A_t(i,j))/\sum_{j=1}^{T_m^l} exp(A_t(i,j)), \tag{11}$$

where $A_t \in R^{T_m^l \times T_m^l}$ is the temporal attention matrix, $\hat{A}_t$ is the normalized (or softmax) attention matrix, the entry $\hat{A}_t(i,j)$ represents the strength of correlation between timeslice i and timeslice j, $X_{T_m^l}^l \cdot \hat{A}_t = \left\{(X_{t-T_m-1}, \ldots, X_{t-1}, X_t)_{l\,layer} \cdot \hat{A}_t\right\} \in R^{N \times F^l \times T_m^l}$ can dynamically adjust the input by merging information in the different time slices of l-th spatial-temporal layer. The parameters $H_1 \in R^N$, $H_2 \in R^{F^l \times N}$, $H_3 \in R^{F^l}$, $V_t \in R^{T_m^l \times T_m^l}$ and $b_t \in R^{T_m^l \times T_m^l}$ are needed to be learned.

In this work, the spatial-temporal attention-based graph wavelet convolution is defined as below.

$$g_\theta * X_{T_m^l}^l = h(((\psi_s \odot S)\Theta^l(\psi_s^{-1} \odot S))X_{T_m^l}^l) \in R^{N \times F^{l+1} \times T_m^{l+1}}, \tag{12}$$

where $*$ indicates the graph wavelet convolution operator, $S = f_{A_s}\left(X_{T_m^l}^l \hat{A}_t\right) \in R^{N \times N}$ is the spatial attention function associated with Eq. (8), Θ^l is the parameterized diagonal matrix, h denotes a non-linear activation function, e.g., ReLU.

Architecture of the Spatial-Temporal Attention Graph Wavelet Network Predictor. In this work, the architecture of the spatial-temporal attention-based graph wavelet convolution network predictor is defined as follows.

Input layer (0-layer):

$$X_0 := X^{t-M:t} \in R^{N \times F \times M}$$

The l-th layer:

$$S = f_{A_s}(X_l f_{A_t}(X_l))$$
$$\hat{X}_l = h(((\boldsymbol{\psi}_s \odot S)\Theta^l(\boldsymbol{\psi}_s^{-1} \odot S))X_l)$$
$$Z_l = ReLU(\hat{X}_l + X_l)$$

The out layer:

$$O_* = UNet\left(\sum_{l=0}^{L} Z_l\right)$$
$$Y = MLP(ReLU(MLP(O_*)))$$

where $ReLU$ denotes the activation function, $UNet$ is the U-Net [20] sub-network, L represents the total number of layers, MLP denotes the multilayer perceptron, $Y \in R^{N \times F \times H}$.

Our proposed predictor stacked multiple spatiotemporal layers to deal with the spatial dependence of different time levels, that is, at the shallowest level, graph convolution receives short-term time information, and at the deepest level, graph convolution processes long-term information. We adopted the mean absolute error (MAE) as the objective function of the model and used the gradient descent method for training our predictor.

4 Experiments

We describe the experimental setting and report the empirical results in this section. In the experiment, the public transportation network datasets PeMSD-4 and PeMSD-8 are used to evaluate the traffic flow predictor.

4.1 Datasets and Settings

We evaluated the STAGWNN on two public highway traffic flow datasets PeMSD-4 and PeMSD-8, which is released by the Caltrans Performance Measurement System (PeMS) [4]. PeMSD-4 records two months of statistics on traffic flow, speed, and occupancy on 307 sensors on the highways of Los Angeles County. PeMSD-8 records two months of statistics on traffic flow, speed, and occupancy on 170 sensors on the highways of Los Angeles County. The readings of the sensors were aggregated into 5-min windows. Each detector detects 12 data points per hour and 288 points per day. Each dataset was partitioned into three sections, which were 60% as training, 20% as validating, and 20% as testing respectively.

All experiments were compiled and tested on Dell Precision 7920 Platform, a Linux Tower Server (CPU: Intel(R) Xeon(R) Gold 5218 CPU @2.30 GHz, GPU: NVIDIA GeForce GTX 2080). In our predictor, all the graph convolution and temporal layers used 64 convolution kernels, $s = \{12.05, 15.05\}$. We adopted the Adam optimizer for parameter optimization with an initial learning rate $lr = 0.0001$. For computational efficiency, we set the elements of ψ_s and ψ_s^{-1} smaller than a threshold $t = \{0.0001, 0.0001\}$ to zero. Meanwhile, we terminated the training if the validation loss does not decrease for 200 consecutive epochs. Furthermore, we randomly shuffled the dataset to obtain the best training effect.

4.2 Metrics and Baselines

Mean absolute errors (MAE), mean absolute percentage errors (MAPE) and root mean squared errors (RMSE) [4] were used as the evaluation metrics. In this work, we compared our traffic flow predictor ASTGWNN with the following baselines:

ARIMA: Auto-Regressive Integrated Moving Average is a time series analysis method for predicting mission. ARIMA is a traditional time series analysis method.
VAR: Vector Auto-Regressive model. VAR is a traditional time series analysis method.
LSTM [12]: Long Short Term Memory network.
GRU [13]: Gated Recurrent Unit network.STGCN(Cheb) [16]: A spatial-temporal graph convolution network based on the spatial method, STGCN(Cheb) with the Chebyshev polynomials approximation.
STGCN(1st) [16]: A spatial-temporal graph convolution model, Chebyshev polynomials order K is 1.
ASTGCN [4]: The attention-based spatial-temporal graph convolution network proposed for the spatial-temporal time series prediction problem.MSTGCN [4]: The multi-component spatial-temporal graph convolution networks which get rid of the spatial-temporal attention.

4.3 Experimental Results

The experimental results are reported in Table 1, which shows the average results of traffic flow forecasting performance over the next hour (12 data points). It can be seen from Table 1 that our predictor, STAGWNN, achieves the best performance with statistical significance in all metrics. We also can observe that the forecasting results of traditional time series analysis methods are usually not ideal, which shows that these methods have limited ability to model nonlinear and complex traffic data. In contrast, the method based on deep learning, such as LSTM, GRU, and STGCN, can usually obtain better prediction results than traditional time series analysis methods. In the Table 1, STGCN(1st), STGCN(Cheb), MSTGCN, and ASTGCN methods can simultaneously capture the temporal and spatial correlations in the time series data.

Table 1. The average performance comparison of different approaches on PeMSD-4 and PeMSD-8.

Model	PeMSD4			PeMSD8		
	MAE	MAPE	RMSE	MAE	MAPE	RMSE
ARIMA	32.15	28.54	68.14	24.16	20.22	43.41
VAR	33.84	26.62	51.87	21.51	19.95	31.53
LSTM	29.45	23.03	45.85	23.37	17.31	36.97
GRU	28.15	22.74	45.42	22.54	16.98	36.01
STGCN(1st)	27.02	21.43	42.14	21.43	16.54	33.97
STGCN(Cheb)	26.89	19.95	39.02	20.27	15.03	30.72
MSTGCN	22.85	17.65	35.82	18.52	14.81	26.85
ASTGCN	21.93	16.86	32.90	16.71	11.06	25.98
STAGWNN (ours)	**21.84**	**16.27**	**32.41**	**16.52**	**10.93**	**25.83**

To investigate the role of fusion U-Net furtherly, we counted the comparison of the performance between STAGWNN with U-Net, and STAGWNN without U-Net (STAGWNN-noUNet) in Table 2. It is easy to observe that our proposal STAGWNN achieves better performance than STAGWNN-noUNet.

Table 2. The average performance comparison between STAGWNN and STAGWNN-noUNet.

Model	PeMSD-4			PeMSD-8		
	MAE	MAPE	RMSE	MAE	MAPE	RMSE
STAGWNN-noUNet (ours)	23.65	17.74	37.09	18.01	12.12	27.94
STAGWNN (ours)	**21.84**	**16.27**	**32.41**	**16.52**	**10.93**	**25.83**

To compare with two methods based on graph wavelet convolution, STAGWNN, and STAGWNN-noUNet, the MAE of the test data set of PeMSD-4 and PeMSD-8 during the training process was plotted in Fig. 4. Those curves also suggest that our model can achieve much faster training procedures and easier convergences. In Fig. 4, we also plot the MAE of ASTGCN. In the first training epoch, the MAE of STAGWNN is 60% lower than the MAE of STAGWNN without UNet both on PeMSD-4 and PeMSD-8. The main reason is that STAGWNN with UNet can simultaneously consider both the spatial-temporal correlations and the complex periodicity of time series data. Hence, our model ASTGWNN can well capture the periodicity of recent-period traffic data, daily-period traffic data, and weekly-period traffic data. It can be seen, in Fig. 4, that the trend of the STAGWNN method and STAGWNN-noUNet method are consistent. This can prove that the model of STAGWNN based on graph wavelet is right in forecasting tasks.

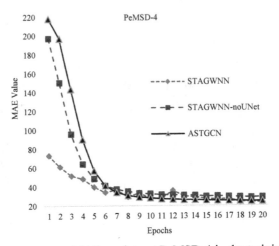

(a) The average test MAE on dataset PeMSD-4 in the training phase.

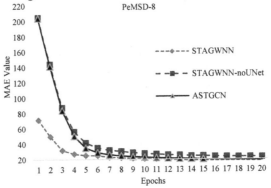

(b) The average test MAE on dataset PeMSD-8 in the training phase.

Fig. 4. Test MAE versus the number of training epochs (PeMSD-4 and PeMSD-8).

Table 3. Statistics of wavelet transform and Fourier transform on PeMSD-4 and PeMSD-8.

Statistical property	Wavelet transform (PeMSD-4/PeMSD-8)		Fourier transform (PeMSD-4/PeMSD-8)	
	ψ_s	ψ_s^{-1}	U	U^{T}
Density	5.77%/21.85%	10.44%/53.09%	99.73%/100%	99.73%/100%
Number of non-zero elements	5439/6314	9835/15344	93997/28900	93997/28900

To demonstrate sparsity in both graph wavelets transform matrix ($\boldsymbol{\psi}_s$ and $\boldsymbol{\psi}_s^{-1}$) and Fourier transform matrix (U and U^T), we conducted the statistical experiment of the sparsity of transform matrixes. We take the PeMSD-4 dataset and PeMSD-8 dataset as examples to illustrate the sparsity of graph wavelet transform. The PeMSD-4 has 307 nodes and the PeMSD-8 has 170 nodes. Hence, the wavelet transform matrix $\boldsymbol{\psi}_s^{-1}$ and the Fourier transform matrix U^T have the same dimensions which belong to $R^{307 \times 307}$ in PeMSD-4 and $R^{170 \times 170}$ in PeMSD-8 respectively. We also display the inverse transform matrix $\boldsymbol{\psi}_s$ and U in Table 3. The first row of the table demonstrates that $\boldsymbol{\psi}_s^{-1}$ is much sparser than U^T both in two datasets. We also find that the inverse wavelet transform matrix is much sparser than U. Therefore, the wavelet transform bases can hold sparsity in the spatial domain.

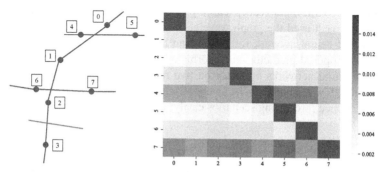

Fig. 5. The attention matrix was obtained from the spatial attention mechanism in a sub-graph with 8 detectors from the PeMSD8.

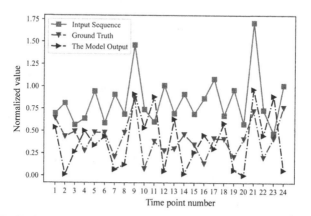

Fig. 6. Traffic flow forecasting in two prediction window of the dataset PeMSD-8.

To investigate the role of attention mechanism in our predictor visually, we conducted a case study: select a sub-graph containing 8 detectors from PeMSD8, and display the average spatial attention matrix between the detectors in the test set. As shown in Fig. 5, in the spatial attention matrix, the ith row represents the correlation strength between each detector and the jth detector. For example, looking at the second line, we can figure out that the traffic on the 2nd detector is closely related to the traffic on the 3rd detector. This is reasonable because the 2nd detector is spatially close to the 3rd detector in the actual traffic network, as shown on the left side of Fig. 5. Furthermore, the 1st detector is offline in this circumstance. Therefore, our graph network-based predictor not only achieves the best predictive performance but also shows the advantage of interpretability. Finally, we showed the predictions of the traffic flow predictor during the two complete forecast windows size as shown in Fig. 6. The results show that the predictions produced by our predictor are very close to the ground truth.

5 Conclusion

We have developed a new graph deep learning predictor for traffic flow forecasting in this paper. Our approach used the graph wavelet diffusion kernels and attention strategies to build a novel neural network for forecasting traffic flow. The experimental results show that the proposed predictor outperforms other state-of-the-art spectral graphs of conventional neural networks predictor on two real-world datasets. Furthermore, our proposed traffic flow predictor can be extended into more general spatiotemporal graph-structured sequence forecasting scenarios easily.

References

1. Abadi, A., Rajabioun, T., Ioannou, P.A.: Traffic flow prediction for road transportation networks with limited traffic data. IEEE Trans. Intell. Transp. Syst. **16**(2), 653–662 (2015)
2. Mackenzie, J., Roddick, J.F., Zito, R.: An evaluation of HTM and LSTM for short-term arterial traffic flow prediction. IEEE Trans. Intell. Transp. Syst. **20**(5), 1847–1857 (2019)
3. Feng, X., Ling, X., Zheng, H., Chen, Z., Xu, Y.: Adaptive multi-kernel SVM with spatial-temporal correlation for short-term traffic flow prediction. IEEE Trans. Intell. Transp. Syst. **20**(6), 2001–2013 (2019)
4. Guo, S., Lin, Y., Feng, N., Song, C., Wan, H.: Attention based spatial-temporal graph convolutional networks for traffic flow forecasting. In: Proceedings of the AAAI Conference on Artificial Intelligence, vol. 33, pp. 922–929 (2019). https://doi.org/10.1609/aaai.v33i01.330 1922
5. Xu, B., Shen, H., Cao, Q., Qiu, Y., Cheng, X.: Graph wavelet neural network. In: International Conference for Learning Representations (ICLR 2019), New Orleans, LA, USA, p. 13 (2019). https://openreview.net/forum?id=H1ewdiR5tQ
6. Donnat, C., Zitnik, M., Hallac, D., Leskovec, J.: Learning structural node embeddings via diffusion wavelets. In: Proceedings of the 24th ACM SIGKDD International Conference on Knowledge Discovery & Data Mining, London, United Kingdom, pp. 1320–1329 (2018). https://doi.org/10.1145/3219819.3220025
7. Hammond, D.K., Vandergheynst, P., Gribonval, R.: Wavelets on graphs via spectral graph theory. Appl. Comput. Harmonic Anal. **30**(2), 129–150 (2011). https://doi.org/10.1016/j.acha. 2010.04.005
8. Tremblay, N., Borgnat, P.: Graph wavelets for multiscale community mining. IEEE Trans. Signal Process. **62**(20), 5227–5239 (2014)
9. Kamarianakis, Y., Vouton, V.: Forecasting traffic flow conditions in an urban network: comparison of multivariate and univariate approaches. Transp. Res. Rec. **1857**, 74–84 (2003). https://doi.org/10.3141/1857-09

10. Williams, B., Hoel, L.: Modeling and forecasting vehicular traffic flow as a seasonal ARIMA process: theoretical basis and empirical results. J. Transp. Eng. **129**, 664–672 (2003). https://doi.org/10.1061/(ASCE)0733-947X(2003)129:6(664)
11. Fradinata, E., Kesuma, Z.M., Rusdiana, S., Zaman, N.: Forecast analysis of instant noodle demand using support vector regression (SVR). In: IOP Conference Series: Materials Science and Engineering, Politeknik Aceh Selatan Campus, Indonesia, vol. 506 (2019). https://doi.org/10.1088/1757-899X/506/1/012021.
12. Ma, Y., Zhang, Z., Ihler, A.: Multi-lane short-term traffic forecasting with convolutional LSTM network. IEEE Access **8**, 34629–34643 (2020)
13. Adege, A.B., Lin, H., Wang, L.: Mobility predictions for IoT devices using gated recurrent unit network. IEEE Internet Things J. **7**(1), 505–517 (2020)
14. Seo, Y., Defferrard, M., Vandergheynst, P., Bresson, X.: Structured sequence modeling with graph convolutional recurrent networks, vol. 1, no. 1, pp. 1–10. arXiv:1612.07659 [stat.ML] (2016)
15. Li, Y., Yu, R., Shahabi, C., Liu, Y.: Diffusion convolutional recurrent neural network: data-driven traffic forecasting. In: International Conference on Learning Representations, Vancouver, BC, Canada, p. 16 (2018)
16. Yu, B., Yin, H., Zhu, Z.: Spatio-temporal graph convolutional networks: a deep learning framework for traffic forecasting. In: Proceedings of the Twenty-Seventh International Joint Conference on Artificial Intelligence, IJCAI 2018, Stockholm, Sweden, 13–19 July 2018, pp. 3634–3640 (2018). https://doi.org/10.24963/ijcai.2018/505.
17. Kipf, T.N., Welling, M.: Semi-supervised classification with graph convolutional networks. In: International Conference on Learning Representations (ICLR 2017), Toulon, France, 24–26 April 2017. https://openreview.net/forum?id=SJU4ayYgl
18. Defferrard, M., Bresson, X., Vandergheynst, P.: Convolutional neural networks on graphs with fast localized spectral filtering. In: Advances in Neural Information Processing Systems 29: Annual Conference on Neural Information Processing Systems 2016, Barcelona, Spain, pp. 3837–3845 (2016)
19. Mishra, K., Basu, S., Maulik, U.: DaNSe: a dilated causal convolutional network based model for load forecasting. In: Deka, B., Maji, P., Mitra, S., Bhattacharyya, D.K., Bora, P.K., Pal, S.K. (eds.) PReMI 2019. LNCS, vol. 11941, pp. 234–241. Springer, Cham (2019). https://doi.org/10.1007/978-3-030-34869-4_26
20. Khanna, A., Londhe, N.D., Gupta, S., Semwal, A.: A deep residual U-Net convolutional neural network for automated lung segmentation in computed tomography images. Biocybern. Biomed. Eng. **40**(3), 1314–1327 (2020). https://doi.org/10.1016/j.bbe.2020.07.007

Vehicle Emergency Route Planning Based on Grid Map

Lei Yuan, Bowen Yang, Yuanying Chi[✉], Zunhao Liu, and Limin Guo

Beijing University of Technology, Beijing 100124, China
{yuanlei,bovin.y}@emails.bjut.edu.cn, {goodcyy,guolimin}@bjut.edu.cn
liuzunhao@outlook.com

Abstract. When an emergency occurs in a city, it causes a large accumulation of vehicles on the roads. It is particularly important to provide effective emergency route planning decisions for vehicles. To make vehicle evacuation more effective and mitigating traffic congestion, a Grid Map Emergency Route Planning (GMERP) methodology is designed. In particular, we first divide the road network into multiple grids and use the connectivity between the grids and the commuting capacity of the road as the weight of the grid. Then, a Grid Map Evacuation Area Recommendation (GM-EAR) method is introduced by considering the road speed, which calculates the commuting capacity of the raster through a sorting algorithm and recommends the raster with stronger computing capacity as the evacuation area. To allow more vehicles to reach the evacuation area in a short time, when planning a path, an Emergency Route Planning Analytic Hierarchy Process (ERP-AHP) method is introduced. The ERP-AHP calculates the road weight by taking into account the factors that affect the road traffic comprehensively so that the planned route has better evacuation ability. The experimental results show that the GMERP model can effectively evacuate vehicles around the congested area.

Keywords: Emergency evacuation · Grid map · Emergency route planning

1 Introduction

Because of the continuous advancement of urbanization in our country, the urban road network has become more complex, the number of private cars increases rapidly [1], the entire urban road network is already heavily loaded with traffic, and vehicle congestion has become an increasingly common phenomenon [2]. When an unconventional event occurs in urban areas, large areas of vehicle congestion will become more severe over time. Planning emergency evacuation routes for vehicles is one of the necessary means to reduce the huge economic losses and casualties caused by emergencies [3]. Therefore, research on emergency navigation has become increasingly important. Emergency navigation should

X. Meng et al. (Eds.): SpatialDI 2020, LNCS 12567, pp. 122–135, 2021.
https://doi.org/10.1007/978-3-030-69873-7_9

ensure the efficiency of vehicles for the evacuation process while reducing the impact and interference on normal traffic.

There is a lot of domestic and international research on emergency evacuation route planning. It mainly uses the dynamic traffic allocation model to select evacuation routes [4]. And the heuristic algorithm is used in the path planning algorithm to give the sub-optimal solution to the evacuation planning problem [5]. Based on the given dynamic road network environment, a co-evolutionary optimization algorithm is proposed to solve the path optimization problem [6]. However, they are mainly qualitative research in the evacuation, as well as quantitative research [7–10]. In addition to this, several researchers have devoted themselves to the study of related evacuation models and evacuation crowding models [11–13].

Due to the complex state of the actual traffic road network and the various road attributes, the above route planning methods did not consider the complex and changeable road network environment. The route planned by these methods is not the most effective evacuation route, which will increase the possibility of secondary congestion during the evacuation process and increase the personnel and economic losses caused by the emergency.

To address these issues, a Grid Map Emergency Route Planning (GMERP) model will be introduced in this paper. This model combines historical trajectory data and various road attributes to plan the optimal emergency route under the current time window. We propose two methods based on GMERP model. The Grid Map Evacuation Area Recommendation (GM-EAR) method calculates the commuting capacity of the road network under different time windows and recommends areas with strong evacuation capabilities to vehicles first. An Emergency Route Planning Analytic Hierarchy Process (ERP-AHP) method is proposed, which integrates multiple attributes of roads to calculate weights and complete emergency route planning. In short, the contributions of this paper are:

(1) This paper proposes a vehicle emergency route planning based on grid maps called GMERP model, which realizes efficient evacuation of vehicles in emergencies;

(2) This paper proposes a method for recommending vehicle evacuation areas under emergency conditions called GM-EAR, which consider the historical trajectory data and interactivity of intersection nodes, rank the importance of all grids in the map, and recommends the best evacuation area;

(3) The ERP-AHP method is proposed to calculate the weight of roads affected by multiple factors and plan the optimal route for vehicles to the evacuation area.

The rest of this paper is organized as follows. Section 2 introduces related work. Section 3 describes the establishment of the model. Section 4 introduces the emergency route planning method based on the grid graph. Section 5 presents the experimental results and evaluation. Finally, conclude this paper in Sect. 6.

2   Related Work

In this section, we will introduce the related work of emergency evacuation and emergency route planning.

So far, regarding the evacuation of vehicles, more research has been carried out by domestic and foreign scholars for obtaining the optimal evacuation route. Sheffi et al. developed an evacuation model that can analyze the characteristics of different traffic modes during the evacuation process [14]. To ensure the optimal evacuation route, Yamada proposed a minimum cost flow model [15]. Brunilde optimized the driving path of the vehicle based on fully considering the reliability of the road network connection [16]. Some scholars regard the shortest evacuation time as the optimal route for vehicles [17,18], without considering the limitations of road commuting capacity, the planned route is likely to cause another congestion.

The purpose of emergency navigation is to first quickly evacuate vehicles in the emergency area to a safe area. Second, reduce the impact of evacuated vehicles on the normal traffic of social vehicles and avoid secondary congestion. Figure 1 shows the basic evacuation rules.

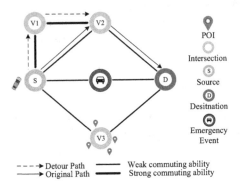

Fig. 1. Basic rules of vehicle evacuation.

The vehicle travels from the source to the destination, and when an emergency occurs nearby. First, areas with unobstructed traffic conditions are used as evacuation areas, and then the road with higher commuting capacity is selected for evacuation.

Evacuation vehicles in an emergency will bring a huge load to the road. When the traffic volume reaches the capacity, the traffic will become unstable. During the planning of an evacuation route, it is necessary to choose a road with a stronger commuting ability, and the commuting ability of a road needs to consider many factors. In summary, the problem is defined as follows: (1) When an emergency occurs on a grid of the grid map, how can nearby vehicles effectively choose the best evacuation area; (2) When calculating the road weight in a grid, how to accurately calculate the road weight under the influence of multiple factors; (3) Existing path planning algorithms cannot perform effective route planning based on grid maps.

3   Model Definitions

3.1   Grid Road Network Model

The grid map is expressed as $Gn = <Pr, Ie, Oe, Grid_{id}>$, where Gn represents the entire $n_g \times n_g$ grid map, Pr indicates a ratio of the speeds of road in the grid to the speeds of road in other grids. The value of this ratio will be considered as the commuting ability of the grid. $Grid_{id}$ represents the ID corresponding to each grid in the grid map, The ID of the grid is composed of r and c, which are the row and column numbers of the grid location. Ie and Oe respectively represent the in-degree and out-degree of each grid. Ie_{r-c} represents the edge set of other grids in the road network pointing to the $Grid_{r-c}$ grid, Oe_{r-c} indicates that the $Grid_{r-c}$ grid pointing to the edge set of other grids in the road network. The GM-REA method can assign values to each grid through these parameters, these values change with the speed of the road network, and also more accurately reflect the change of the grid weight.

3.2   Grid Map Emergency Route Planning Model

The dynamic road network can be defined by a directed graph $R = <G, T>$. Where T represents the time cycle of the road network update. In a cycle, the road network can be considered as a static directed graph G, defined by $G = <V, E, Rs, W>$. Where V is the set of intersection nodes in the road network, and $v_i \in V$ is one of the nodes. E is the set of all edges, $e_{ij} \in E$ represents the path from node v_i to node v_j. Noted that the direction of each path is a vector and the time at point v_i is ahead of v_j time. With the change of the T, directed graph G has different road network states $G^1, G^2, ..., G^T$. Rs is the set of road speeds in the road network, expressed as $Rs = \left\{ Rs^1, Rs^2, ..., Rs^t \right\} = \left\{ \{rs_1^1, rs_2^1, ..., rs_L^1\}, ..., \{rs_1^t, rs_2^t, ..., rs_L^t\} \right\}, t \in T$. Where rs_L^t represents the road speed of the path L between adjacent nodes in the t-th time cycle. The weight set of the road is $W = <v_i, v_{i+1}, W_L>$, W_L is defined as the road weight between the adjacent node v_i to the node v_{i+1}. W_L is a non-negative number is all lengths are positive. Different from the commonly used $Euclidean$ distance measurement, The value of W_L is restricted by many factors, W_L can be expressed as:

$$W_L = (\frac{l_L}{rs_L^t})^{-1} + (\alpha_1 \omega_1 + \alpha_2 \omega_2 + ... + \alpha_m \omega_m) \tag{1}$$

Where $\frac{l_L}{rs_L^t}$ represents the time spent passing path L at t, l_L is the length of L. The weight factors ω_m represents m factors that affect the weight of the road. α is the corresponding weighting coefficient, and the weight coefficient of each factor is calculated by the ERP-AHP method. When calculating road weight coefficients, establish target layer, criterion layer, sub-criteria layer, and scheme layer. The goal layer is denoted as Z, the criterion layer is composed of m factors that influence the goal, expressed as $C_m = c_1, c_2, \ldots, c_m$. The sub-criteria layer contains the influencing factors $SC_k = sc_1, sc_2, \ldots, sc_k$ for each factor of the criterion layer, where k is the number of factors in the sub-criteria layer. When there are n edges in the road network, and the scheme layer is composed of n edges, expressed as $E_n = edge_1, edge_2, \ldots, edge_n$. The factors of the criterion layer are compared in pairs to construct a $m \times m$ judgment matrix A. The element a_{ij} of matrix A represents the relative importance of the i-th criterion to the j-th criterion. The eigenvector of matrix $\alpha = (\alpha_1, \alpha_2, \ldots, \alpha_m)^T$ is the weight coefficient of each factor. W_L is calculated according to the road attributes and the corresponding influence coefficient. Based on the road weight, the emergency route planning can be defined as a function $max(W_{L1}, W_{L2}, \ldots)$ on the continuous edge, by finding the path with the largest road weight as the optimal solution.

4   Method

4.1   Evacuation Area Recommendation Based on GM-ERA Method

According to our previous research and work [19]. Figure 2 shows a grid map of Beijing's road network within the fifth ring after the grid into $n_g = 144$ and $n_g = 288$.

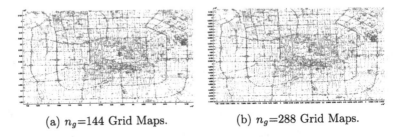

(a) n_g=144 Grid Maps.   (b) n_g=288 Grid Maps.

Fig. 2. Differently distributed grid maps within the fifth ring.

When an emergency occurs, the first response of the GMERP module is to combine road network data and historical trajectory data to recommend the most suitable evacuation area to the user based on the road conditions at the time of the emergency. The number of intersection nodes in each grid, in-degree, out-degree, and road network speed is used as the standard to measure the evacuation capacity of the grid (i.e. commuting capacity). The following introduces a GM-ERA method, which calculates and ranks the commuting capacity of the grid.

The grid map's purpose is to map the entire city to different grids and calculate the number of intersection nodes and the speed of the roads as the weight of the grid. In the GM-ERA method, the grid sorting recommendation algorithm is an improved Dynamic Grid PageRank (DGPR) algorithm based on the traditional PageRank algorithm. Page et al. proposed the traditional PageRank algorithm [20]. The core formula of the algorithm is Formula (2).

$$P(w) = d \sum \frac{P(m)}{N_m} + \frac{(1-d)}{N} \tag{2}$$

Where $P(w)$ represents the ranking value of the web page, d is the damping coefficient, m is the current web page, $P(m)$ indicate the rank value of the current page m, N_m is the number of links of the page, and N is the total number of web pages.

The linked web pages can be thought to be a large directed graph, which is consistent with the abstract directed graph of the grid road network in this paper. Figure 3 shows a schematic diagram of the grid road network, which divides the road network into 2×2 grids. The blue points in the figure represent the road nodes in the road network, and the red lines represent the roads that are connected between different grids. A grid can be regarded as a web page, and the interconnecting roads between the grids can be regarded as links between web pages.

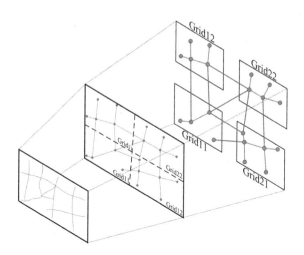

Fig. 3. The schematic diagram of the grid road network.

Assuming that in a time cycle, the speed of the road is constant. In the Fig. 3, $Grid_{11}$ contains four nodes, and $Grid_{22}$ contains five nodes, then it is considered that the commuting capacity of $Grid_{11}$ is stronger than that of $Grid_{22}$ in terms of the number of nodes. However, the actual situation also needs to consider road connectivity and road speed when calculating the grid commuting capacity.

The speed of the road is denoted as Rs, and Rs are critical in the sorting of the grid. The rank value of each grid can be calculated by Formula (3) and Formula (4).

$$DGPR(Grid_{\mathrm{id}}) = \sum_{i=1}^{k \in N} Pr_i * (V_{Grid_{id}}/V) + (1-d)/N \qquad (3)$$

$$Pr_i = \sum_{j=1}^{Vc} Ie_j * d * (Rs_j/120) \qquad (4)$$

Where $DGPR(Grid_{id})$ is the rank value of the $Grid_{id}$ after iterating through the DGPR algorithm. This value represents the commuting capacity of the grid. k is the number of grids directly connected to the node in the map, N represents the number of grids containing nodes. Pr_i is the DGPR value of i-th grid, i stands for the ID of the grid. $V_{Grid_{id}}$ is a total number of vertices in $Grid_{id}$, and V is vertices total number in the whole network. The damping coefficient d is 0.85. Ie_j represents the in-degree of the j-th grid. $(Rs/120)$ represents the normalization of road speed.

Algorithm 1. DGPR

1: **Input:** road network G^T , trajectory data and update time step t
2: **Output:** $Grid_{id}$, Pr_i, $ERA(Grid_{id})$
3: **for** coordinate in G **do**
4:    $Grid_{id}$ = ID
5:    **for** calculation number of grid and edge **do**
6:       edge = weight $< v_i, v_j >$
7:       A_{matrix} = Edge $outdegree < v_i, v_j >$
8:    **end for**
9: **end for**
10: **Initialization** $Pr_i=1/N$
11: **for** Initialization $(Grid_{id})$ **do**
12:    $Pr_0=1/n$
13:    $k=1$
14:    **if** emergencies == 1 **then**
15:       **for** G **do**
16:          $Rs^t = rs^t_L < v_i, v_j >$
17:       **end for**
18:       Cg_{ID} = Unsafe $Grid_{id}$
19:       **for** $Pr_{k+1}=A_{matrix} * e * Pi * (Rs/120) + (1-d)/N$ **do**
20:          k=k+1
21:          **if** $Pr_{k+1}-Pr_k<$sum **then**
22:             **Return** $Grid_{id},Pr_i$
23:          **end if**
24:       **end for**
25:       **for** Search the $Grid_{id}<ID_1,ID_2,\ldots,ID_n>$near Cg_{ID} **do**
26:          $ERA(Grid_{id}) = Grid_{id}$ have max value of R_i
27:          **Return** $ERA(Grid_{id})$
28:       **end for**
29:    **end if**
30: **end for**

The DGPR algorithm first loads the road network data and historical trajectory data and sets the road network update step t. Mapping the road network into a two-dimensional coordinate system and divide it into grids. Assign a unique ID to each grid, and count the Ie and Oe of each grid. Initialized Pr_i = 1, calculate $DGPR(Grid_{id})$. Take t as the step size, dynamically update the road speed and $DGPR(Grid_{id})$. When an emergency occurs on a grid, get the ID of that grid, and the rank value of the grid is calculated with the latest road network speed. Finally, the algorithm returns the grid with the highest value.

4.2   Emergency Route Planning Based on ERP-AHP Method

The ERP-AHP method first determines the weight of the road through the analytic hierarchy process, then uses the improved Bidirectional Dijkstra to plan the emergency route, which is called Dynamic Grid Bidirectional Dijkstra (DGBD) algorithm.

As shown in Fig. 4, when calculating road weights, it is necessary to consider the impact of *Road Width*, *Road Length* and *Road Grade* on road commuting capacity. The above three factors that affect the weight of the road are used as criteria to form the criteria layer.

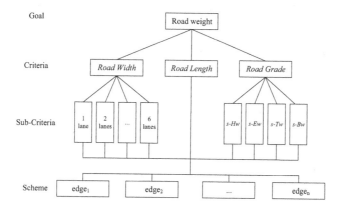

Fig. 4. The hierarchical structure of the ERP-AHP method.

In the criterion level, for the length of the road, the longer the road during an evacuation can accommodate more vehicles. The width of the road directly affects the commuting capacity of the road, and the number of lanes in the road network data can determine the width of the road. The more lanes there are, when the traffic volume increases, there will be a better commuting ability. The roads are divided into highway, expressways, trunk, and branch, they are expressed as $s - Hw$, $s - Ew$, $s - Tw$, and $s - Bw$.

According to the relative weights determined between factors, the judgment matrix A is constructed. Calculate the eigenvector α and maximum eigenvalue

λmax of the matrix by Formula (5):

$$A\alpha = \lambda_{max}\alpha, \alpha = (\alpha_1, \alpha_2, ...\alpha_m)^T \tag{5}$$

Where α_m corresponds to the weight coefficient of the m-th factor in the criterion layer. Finally calculating the consistency index CI and the consistency ratio CR by Formula (6) and Formula (7) respectively:

$$CI = (\lambda_{max} - n)/(n - 1) \tag{6}$$

$$CR = CI/RI \tag{7}$$

Among them, the closer the value of CI is to 0, the higher the degree of consistency. The value of CR is less than 0.1 to pass the consistency test. In this way, the eigenvector α can be used as the weight vector(the value of the weight vector will be given in detail in the experimental section).

The path planning algorithm based on the grid road network not only considers the weight of the road but also plans the evacuation route according to the connection relationship between the grid. The general path planning algorithm is no longer suitable for the current situation. The DGBD algorithm judges the relative position relationship between the vehicle and the emergency in the grid map, guides the vehicle to a strong evacuation capacity grid, and then plans the evacuation route according to the weight of the road.

Algorithm 2. DGBD

1: **Input:** G_n, *string point* v_s, *ending point* v_d, t
2: **Output:** N_s, G_s, T
3: **Initialization:** $W_L[i]$, Path$[i]$, $S_{qu} = \{v_s\}$, $D_{qu} = \{v_d\}$, Time $= t$
4: **for** search $v_s.next$ and $v_s.next$ **do**
5:     **if** $v_s.Grid_{id} == v_d.Grid_{id}$ **then**
6:         **if** $v_s.nxext != v_d.nxext$ **then**
7:             $S_{qu}.add(Max|W_L[v_s.next])$ and $D_{qu}.add(Max|W_L[v_d.next])$
8:             **else** $N_s = S_{qu} + D_{qu}$ and $T = \sum_L^{N_s} \frac{l_L}{rs_L^t}$
9:             **Return** N_s, T and **break**
10:        **end if**
11:    **else for** search $v_s.Grid_{id}$ and $v_d.Grid_{id}$
12:        $G_s = v_i.Grid_{id} + v_j.Grid_{id}$ which $Grid_{id}$ have $Max|DGPR(Grid_{id})$
13:        **Return** G_s and **break**
14:    **end for**
15: **end if**
16: **end for**

The DGBD algorithm process is described as follows. Get the current node of the vehicle, and judge whether the next search node and the current node is in the same grid. If two nodes are in the same grid, the optimal road path is planned according to the road weight. When it is not in the same grid, the algorithm preferentially selects the node of the grid with the highest ranking as the next search node. When the search node is in the recommended evacuation area, the algorithm stops and returns the evacuation route and travel time T.

5 Experiment and Result Analysis

5.1 Experiment Settings

Due to the following two reasons, the experiment in this paper is not verified against the existing route planning methods or algorithms: (1) The data sets used in the experiment are different. This experiment uses Beijing road network data and Beijing historical trajectory data. Different road network data sets have different attributes, and the sampling frequency in historical trajectory data is different. (2) Different from the traditional road network structure, this experiment is based on the grid road network for route planning.

In the actual traffic network, the carrying capacity of the road is constant. And follow the first-in-first-out principle in the driving process of the vehicle. We will conduct two experiments in this paper under this rule.

1. Calculate the commuting capacity of the grid based on the grid map and historical trajectory data, and sort the grid.
2. The comparative test of vehicle travel time under different weight coefficients verifies the validity of the model.

The experiment was performed on a PC with 64-bit Windows, Intel i5 @2.30 GHz, and 16 GB RAM through Java. The dataset includes historical trajectory data in Beijing and Beijing road network data from OpenStreetMap.

5.2 Calculate the Grid Commuting Capacity and Rank

The experiments are based on the grid maps $n_g = 144$ and $n_g = 288$ shown in Fig. 2. According to Formula (3) and Formula (4), calculate the commuting capacity of each grid containing nodes. Count the number of nodes contained in each grid and the number of in-degree edges. Assign 1 to the initial commuting capacity value of each grid. For example, when a grid contains 5 in-degree edges, the initial weight of each edge is $1/5$. In the algorithm iteration, the normalized road speed is re-assigned for each edge until the rank value converges. The road network updates the speed value of the road after the time interval of the step t and recalculates the raster rank value and sorts it. Table 1 shows the commuting capacity values and ranking results of grids with $n_g = 144$ and $n_g = 288$.

Table 1. The commuting capacity values and ranking results of the grid.

$n_g = 144$			$n_g = 288$		
Rank	$Grid_{id}$	Value	Rank	$Grid_{id}$	Value
1	80-65	0.0015708543	1	28-91	0.00045902558
2	38-29	0.0013962760	2	229-123	0.00040509173
3	44-94	0.0013588988	3	77-34	0.00040352772
4	47-47	0.0013328366	4	108-63	0.00039542890
5	42-47	0.0013319262	5	161-50	0.00037715083
6	98-37	0.0012617575	6	160-35	0.00037070340
7	27-90	0.0012450919	7	175-173	0.00036740946
8	79-112	0.0012118589	8	167-44	0.00036339510
9	104-32	0.0011866940	9	174-94	0.00036333606
10	39-32	0.0011768892	10	170-214	0.00036276187

Table 1 shows the sorting results of the grid in descending order (the query time is 2 pm). Where $Rank$ is the rank of the grid, and $Value$ represents the commuting capacity of the grid calculated by the DGPR algorithm. For example, when $n_g = 144$, $Grid_{80-65}$ with the strongest commuting capacity, $n_g = 288$, $Grid_{28-91}$ with the strongest commuting capacity.

When an emergency occurs, nearby vehicles need to be evacuated in an emergency. The commuting capacity of the grid reflects the importance of the grid in the road network. The larger the value of the grid, the stronger its ability to dredge vehicles in the road network. Therefore, if the grid with a higher rank value is used as the evacuation area, the navigation has a better evacuation effect.

5.3   The Effectiveness of the GMERP Modle

This section is a comparative experiment of vehicle evacuation efficiency under different weight coefficients of the GMERP model. Set the relative importance between the road weight factors in The ERP-AHP method, and obtain the following four judgment matrices:

$$A1 = \begin{bmatrix} 1 & 7 & 2 \\ 1/7 & 1 & 1/8 \\ 1/2 & 8 & 1 \end{bmatrix} \quad A2 = \begin{bmatrix} 1 & 2 & 5 \\ 1/2 & 1 & 4 \\ 1/5 & 1/4 & 1 \end{bmatrix} \quad A3 = \begin{bmatrix} 1 & 1/3 & 1/5 \\ 3 & 1 & 1/2 \\ 5 & 2 & 1 \end{bmatrix} \quad A4 = \begin{bmatrix} 1 & 1/5 & 1/2 \\ 5 & 1 & 3 \\ 2 & 1/3 & 1 \end{bmatrix}$$

When the number of factors in the criterion layer is 3, $RI = 0.58$. The eigenvectors α and CI and CR of the four matrices are calculated by Formula (5), (6), and (7) as: $\alpha 1 = (0.5621, 0.0632, 0.3748)^T$, $CI = 0.03854$, $CR = 0.0665$. $\alpha 2 = (0.5679, 0.3334, 0.0982)^T$, $CI = 0.01232$, $CR = 0.0213$. $\alpha 3 = (0.1095, 0.3092, 0.5479)^T$, $CI = 0.00185$, $CR = 0.0031$. $\alpha 4 =$

$(0.1221, 0.6479, 0.2298)^T$, $CI = 0.0018$, $CR = 0.0032$. All pass the consistency test.

Set up five comparative experiments. Put the above four groups of weight coefficient into Formula (1), calculate the weight of the road in the road network. The weight vector of the fifth experiment is $\alpha5$, and only the road length is considered when calculating the road weight. Set the query time at 8:30 am and 2:00 pm to simulate the congested road network during rush hour and the unobstructed road network during off peak, the speed is assigned to the road based on the query time and historical trajectory data. Set the same starting point, ending point and query time in the comparative experiment. In the experiment, we set the starting point as the west gate of Beijing University of Technology, and the ending point as the Institute of Software, Chinese Academy of Sciences. The experimental results as Table 2:

Table 2. Comparison of the effect of querying multiple paths under different time period.

Case	W_L	Str_t	$Dis\,(km)$	$S_{avg}\,(km/h)$	$Tr_t\,(min)$
Off peak	$\alpha1$	14:00	26.9	47.44	34
	$\alpha2$	14:00	26.5	49.72	32
	$\alpha3$	14:00	23.0	46.00	30
	$\alpha4$	14:00	25.7	49.71	31
	$\alpha5$	14:00	22.1	28.22	47
Rush hour	$\alpha1$	8:30	26.9	19.21	84
	$\alpha2$	8:30	**26.5**	**20.08**	**79**
	$\alpha3$	8:30	25.7	18.57	83
	$\alpha4$	8:30	23.4	16.36	86
	$\alpha5$	8:30	22.1	12.85	103

Table 2 shows the results of the path distance (Dis), average speed (S_{avg}), and travel time (Tr_t) planned by the algorithm when the road weight coefficient takes different values when the model is at rush hour and off peak.

During off peak, we can see that under the influence of different weight coefficients, although the planned routes are different because the GMERA method adds the calculation of the road network speed when recommending the area. Although the $\alpha5$ has an advantage in the distance, it does not have an advantage in travel time due to road speed limitations. In the other four cases, even in the case of unobstructed road conditions, the choice of roads still follows the priority of commuting ability, and the vehicle guidance effect is better. When in rush hour, In terms of driving distance, when the algorithm takes the path length as the road weight, the planned route length is the shortest. However, because the road conditions are not considered in this case, the driving time is the longest. For the Tr_t, although the proportion of the path length in the weight calculation decreases when the weight coefficient is $\alpha2$, resulting in a 19% increase in the distance traveled, the travel time is shortened by 35%. In this case, the S_{avg} of the vehicle is also the fastest. For vehicle evacuation, there is a better evacuation effect when the weight coefficient is $\alpha2$.

Through the above comparison experiment, it is proved that in the case of choosing the appropriate weight factor, combined with the evacuation area recommended by the GM-ERA method, the ERP-AHP method can effectively plan to avoid congested roads and choose roads with strong traffic capacity. In an emergency, the GMERP model can effectively evacuate vehicles even if there is a large amount of congestion on the road.

6   Conclusion

In summary, our proposed GMERP model based on grid maps can effectively evacuate vehicles. The GM-ERA method in the model realizes the calculation and sorting of the commuting ability of the grid map and recommends a grid with a stronger evacuation ability to vehicles in the dangerous area, avoiding the secondary congestion problem caused by poor commuting in the evacuation area. After recommending evacuation areas for vehicles, the ERP-AHP method integrates multiple attributes of the road network to calculate road weights and plan evacuation routes for vehicles, thereby effectively improving evacuation efficiency. Experiments show that the GMERP model has a good effect on the evacuation of vehicles in emergencies.

In future research, we will consider more factors that affect the road network, such as POI density and weather. Therefore, a more comprehensive emergency navigation model is our research focus in future research.

Acknowledgment. This work is supported by National Key R&D Program of China (No. 2017YFC0803300), the Beijing Natural Science Foundation (No. 4192004), the National Natural Science of Foundation of China (No. 61703013, 91646201), the Project of Beijing Municipal Education Commission (No. KM201810005023, KM201810005024).

References

1. He, Z., Qi, G., Lu, L.: Network-wide identification of turn-level intersection congestion using only low-frequency probe vehicle data. Transp. Res. Part C Emerg. Technol. **108**, 320–339 (2019)
2. Yang, Z.H.: Analysis of the impacts of open residential communities on road traffic based on AHP and fuzzy theory. Ingénierie des Systèmes dInformation **25**(2), 183–190 (2020)
3. Sharma, S.: Simulation and modeling of group behavior during emergency evacuation. In: IEEE Symposium on Intelligent Agents. IEEE (2009)
4. Mahmassani, H.S.: Dynamic network traffic assignment and simulation methodology for advanced system management applications. Netw. Spatial Econ. **1**, 267–292 (2001)

5. Lu, Q., George, B., Shekhar, S.: Capacity constrained routing algorithms for evacuation planning: a summary of results. In: Bauzer Medeiros, C., Egenhofer, M.J., Bertino, E. (eds.) SSTD 2005. LNCS, vol. 3633, pp. 291–307. Springer, Heidelberg (2005). https://doi.org/10.1007/11535331_17
6. Hu, X.B., Zhang, M.K., Zhang, Q.: Co-evolutionary path optimization by ripple-spreading algorithm. Transp. Res. Part B Methodol. **106**, 4535–4542 (2017)
7. Chun-Hui D, Science S O, University N M.: Urban traffic emergency evacuation route optimization simulation. Comput. Simul. (2017)
8. Giovanna, C., Giuseppe, M., Antonio, P.: Transport models and intelligent transportation system to support urban evacuation planning process. IET Intell. Transp. Syst. **10**, 279–286 (2016)
9. Konstantinidou, M.A., Kepaptsoglou, K.L., Karlaftis, M.G.: Joint evacuation and emergency traffic management model with consideration of emergency response needs. Transp. Res. Rec. J. Transp. Res. Board **2532**, 107–117 (2015)
10. Lv, Y., Zhang, X., Kang, W.: Managing emergency traffic evacuation with a partially random destination allocation strategy: a computational-experiment-based optimization approach. IEEE Trans. Intell. Transp. Syst. **16**, 2182–2191 (2015)
11. Yamabe, S., Hasegawa, F., Suzuki, T.: Driver behavior response to information presentation based on the emergency evacuation procedure of the great east japan earthquake. Int. J. Intell. Transp. Syst. Res. **17**, 223–231 (2019)
12. Milenkovic, M., Kekic, D.: Using GIS in emergency management. In: International Scientific Conference on ICT and E-Business Related Research (2016)
13. Lochhead, I., Hedley, N.: Mixed reality emergency management: bringing virtual evacuation simulations into real-world built environments. Int. J. Digit. Earth **12**, 1–19 (2018)
14. Sheffi, Y.: A transportation network evacuation model. Transp. Res. Part A Gen. **16**, 209–218 (1982)
15. Yamada, T.: A network flow approach to a city emergency evacuation planning. Int. J. Syst. Sci. **27**, 931–936 (1996)
16. Brunilde, S., Soumis, F.: Communication and transportation network reliability using routing models. IEEE Trans. Reliab. **40**, 29–38 (1991)
17. Jia S, Wen-Mei G.: Selection of optimal emergency logistics path under a time-varying condition. China Saf. Sci. J. (2015)
18. Yuan, Z., Xue-Qiang, W., Guo-Ming, C.: Evacuating route optimization based on the minimum toxic dose in toxic gas-leaking accidents. J. Saf. Environ. **13**, 266–270 (2013)
19. Yang, B., Ding, Z., Yuan, L.: A novel urban emergency path planning method based on vector grid map. IEEE Access **9**, 338–353 (2020)
20. L. Page, B. Sergey, M. Rajeev, and W. Terry.: The PageRank Citation Ranking: Bringing Order to the Web, Stanford InfoLab, pp. 1–14 (1999)

Data Science

Application of Multi-agent System in Land Use Decision-Making of Industrial Park: A Case Study of Tianfo Health Industrial Park, Shandong

Wenhan Feng[1]([✉]) [iD], Peiyao Song[2] [iD], and Bayi Li[3] [iD]

[1] RWTH Aachen University, Aachen 52062, Germany
feng_wenhan@126.com
[2] Chongqing University, Chongqing 400030, China
[3] Shanghai University, Shanghai 200444, China

Abstract. Exploring the driving mechanism of land-use distribution is one of the foundations to ensure sustainable urban development. Industrial parks have great significance in coordinating industrial land with other types of land use for human settlements and economic development. This paper examines the development decisions of industrial parks from the perspective of complex systems by constructing a general framework for simulating the dynamics where develop a multi-agent model (LUL-IV, Land Use Layout – Industrial Vision) based on the NetLogo Platform. The research combines government land-use control policy from the macro perspective with the developer's decision-making behavior from the micro perspective. Based on the evaluation of land-use suitability, we set the different behavioral rules for different subjects according to the demand for living space. In this study, the model defines three agents, including government, developers and residents, to generate the land-use distribution of Tianfo Health Industrial Park in Shandong Province. Furthermore, the results can be used as a tool to support planning decisions.

Keywords: Land use · Dynamics simulation · Agent-based modeling · Industrial park · Decision-making

1 Introduction

Geosimulation has been widely adopted to support the decision-making process of sustainable land use in urban planning projects. Due to the advantages in simulating discrete and dynamic geospatial object sets, automation-based simulation has become the mainstream method of urban geographic simulation research after a long period of development. The general techniques include Cellular Automata (CA) and Agent-based Model (ABM). Once setting the initial state, CA is automated and unable to be modified accordingly. Its highly simplified model feature is suitable for the deduction of low precision in macroscale. By comparison, ABM can effectively simulate the dynamic decision-making process of land development in multi-level and has more potential in studying the complexity of human spatial behavior [1]. Therefore, ABM is more flexible and systematic.

More studies of geosimulation in recent years [2] focus on the ABM method and its applications. Neural networks and ABM were discussed in the application of the urban land-use change modeling, then ABM was applied to the planning of industrial parks, and a basic framework for the development of creative parks was also put forward in 2013 [3]. ABM shows great potential in spatial behavior analysis. A behavior analysis framework for enterprise migration based on the ABM method is adopted to explore the mechanism of decision-making [4]. The decision behaviors of the stakeholders and corresponding spatial changes in ABM modeling were further discussed, with stressing the role of experience [5]. The application of ABM is developed by taking the comprehensive factors of urban society into account [6]. Simulating the spatial changes of the real estate market by ABM makes preparation for future planning, which can timely respond to the evolution of residential land in urban development [7].

The urban land system, as a dynamic and complex system, is constantly updating and changing. These spatial changes are influenced by the government policy and the demands for interest at the micro- and macro-level [9]. A successfully planned industrial park can boost urban economic growth and industrial development, moreover, fulfilling the needs of the stakeholders in the city. China's government is also trying to develop industrial parks in order to become more competitive and sustainable. The study explores the health industrial development strategy, including the behavioral rules of different subjects as well as the relationship between control and autonomy in urban development. We take both macro factors and micro factors into consideration in the ABM simulation, trying to layout the land use planning of Tianfo Health Industrial Park in Shandong province so as to better balance the interests and demands of multiple subjects.

In this study, we demonstrate a three-level generalized framework, including an environmental factors layer, an action rule layer, and a decision-making layer. Under this framework, the action rules and schemes of three representative agents: developers, residents, and the government are discussed. Applying these rules of social behavior and decision-making to the multi-agent model, urban spatial changes influenced by these three agents can be simulated. The model in the study, LUL-IV (Land Use Layout – Industrial Vision), was developed based on the NetLogo platform. This model is designed for the simulation of land use development in industrial parks. It reflects the dynamical land-use changes and the gaming process of three agents, simulating the spatial and quantitative relationship among various land-use types. The feasibility is verified by applying the system to the development planning of the Tianfo Industrial Park in Linyi City, Shandong Province.

2   The General Framework for Simulating the Dynamics

The model framework is the apriority and assumptions of the simulation system in certain conditions. As a result, model operators can develop a better understanding and controllability of the simulation process [8]. The framework is built based on bounded rational decision-making, which is composed of three layers: the layer of environmental factors, the layer of the multi-agent behavior rule, and the layer of behavior decision-making (Fig. 1).

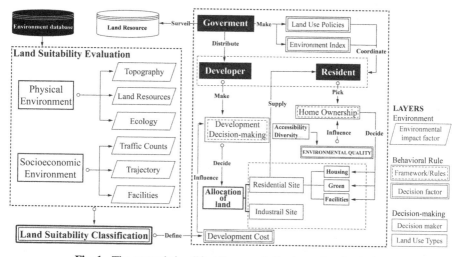

Fig. 1. The general simplified framework for dynamics simulation

2.1 Environmental Factors Layer

In this layer, both the natural environment and socio-economic environment are seen as the environmental factors. They define the space in which agents operate and interact with the environment and each other. The natural environment includes factors such as topography, land resources, and ecological environment, while the socio-economic environment includes factors such as population, traffic, and public facilities. The features of these environmental factors are usually indicated in geographic data and statistical data. The preprocessing of these data provides an environmental basis for a more indepth analysis of the framework and a stage of agent's action and decision. Through the analysis of the present environment, we can obtain the standpoint of different action subjects. Therefore, the environmental factor layer is the propelling force of the simulation in the model.

2.2 Action Rule Layer

The layer of action rule defines the general rules of gaming, constraining agent's moves in a specific framework. According to the features of the development process of industrial parks, three representative agents: developers, residents, and government are considered in the framework. For developers, the main concern is the development cost, that is, how to achieve more economic benefits significantly. For residents, they focus on the quality of their living environment, which is decided by functions such as accessibility to green space and public facilities. Besides, the government's responsibilities are to regulate the land development intensity at a macroscopic level and to balance the interest of the other two agents. In the system, three agents will take strategies to iterate and game, following the action rules. To reach an optimal ratio of land use, residents and developers will restrict each other under the government's regulation. Eventually, the land-use layout would become stable and acceptable.

The action rules of agents in the gaming process as follow:

- Residents and developers as initiatives start the game which is under government's regulations;
- Residents and developers meet bounded rational decision-making hypothesis, and pursue the maximization of their interests separately;
- The interests and objectives of each agent are different.

2.3 Decision-Making Layer

The decision-making layer is the calculation and iteration core of the model. Specifically, it includes three sub-processes: location decision of developers, location decision of residents, and planning regulation of the government.

Location Decision of Developers. According to the theory of bounded rational decision-making, developers mainly give priority to the area where generates the maximum economic benefits. Specifically, these areas are usually with less difficulty in construction, convenient transportation, and fewer built entities [16]. With the change of land use pattern in the process of development, developers prefer to develop the land which is conducive to industrial development.

Generally speaking, developers will optimize the allocation of land resources and promote the full benefits considering the situation of the physical environment, socio-economic, and ecology. However, the actions of developers should also be regulated by the government. According to the Code for Vertical Planning on Urban and Rural Development Land, the land with a natural slope of less than 15% is suitable for industrial storage, public construction, and residential planning. The area with a natural slope between 15% and 25% can only be developed for residential and public construction. The land with a natural slope of more than 25% should not be carried out the construction. In summary, developers evaluate land-use suitability by selecting and weighting physical and socio-economic factors to obtain the optimal layout plan of industrial land under the framework.

Location Decision of Residents. The concern of residents is related to housing, such as surrounding green space and public service facilities. To reach the best allocation, residents need to make empirical judgments based on their demand for living. After the comprehensive comparison, we used the concept of the 15-min neighborhoods proposed in the Standard for Urban Residential Planning and Design GB50180-2018. As the basic unit for community life, the 15-min neighborhoods should satisfy the needs of all kinds of daily activities, and surrounding public service facilities should be accessible within 15 min of walking. The radius of a 15-min neighborhood is about 800 m, typically equal to a 15-min walking distance, which ensures an environment-friendly and multi-functional community [11, 14, 15]. Residents in the model make their own choice for housing location in accordance with the residential land control indicators specified in the Residential Land Control Index of 15-min Pedestrian-scale Neighborhood (Table 1). Within a 15-min neighborhood, the residential area constitutes 48%–61%, green space constitutes 7%–16%, and public service facilities area constitutes 12%–23%. Public service facilities and green space are the key factors in improving the living quality of residents. To reach a maximum utility, the choice for housing location is mainly based on the conditions of the environmental factors layer by checking the accessibility to green space and public service facilities.

Table 1. Residential land control index of 15-min pedestrian-scale neighborhood (standard for urban residential planning and design GB50180-2018).

Housing category	Composition of residential land (%)			
	Housing	Public facilities	Green	Road
Multi-storey I (4–6 stories)	58–61	12–16	7–11	15–20
Multi-storey II (4–6 stories)	52–58	13–16	9–13	15–20
Highrise (4–6 stories)	48–52	16–23	11–16	15–20

Government Planning Regulation. The role which the government plays in urban land use development is both macro-regulator and participant. Developers and residents are gaming for the maximization of their benefits, while the government needs to regulate both of them based on the theory of bounded rational decision-making [13]. Practically, the government should consider the general interests rather than individual interests. The regulation will occur in the following situations:

- When the proportion of industrial land is too high, and the development intensity exceeds the limit of land development;
- When the allocation of land use within the 15-min neighborhoods exceeds a certain percentage, which is unacceptable to residents;
- Excessive gathering of residential land, which cannot ensure the interests of other agents in the area;

In these cases, the government will act to re-plan, so that developers and residents could restart the game and adjust the proportion of land use until reaching a balanced state.

3   Land Use Decision-Making System

3.1   Research Methods

The whole framework consists of two parts: data preprocessing and agent-based model (LUL-IV) (Fig. 2).

Data preprocessing includes analyzing factors that influence land use in industrial parks and establishing the index system for evaluating the suitability of land development. ArcGIS 10.2 is used to perform spatial analysis of original data on traffic conditions, built entities, landscapes, and environmental quality. Then the influence degree of each factor is calculated and weighted based on the development suitability of industrial land.

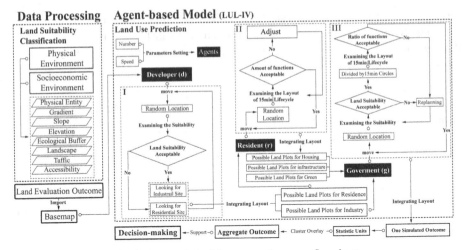

Fig. 2. Research decision tree and process flowchart

Finally, dividing the land-use suitability level according to the weighted score of the suitability, which supports the decision-making of the developers in the ABM.

The agent-based model LUL-IV (Land Use Layout - Industrial Vision) is a land-use layout evolutionary model for industrial park planning. This model is developed based on the multi-agent modeling platform NetLogo, the interface of which is user-friendly and easy to operate. NetLogo provides three basic types of agents: turtles, patches, and links. Turtle agents can move freely, while patch agents that are similar to cells in the cellular automaton are limited to move [12]. Three types of agents are defined with the turtle agents: developer (d), resident (r), and government (g), and visualize the process of land-use change by different colors.

3.2   Agent-Based Model LUL-IV (Land Use Layout – Industrial Vision)

The Dynamics System. In the model, the change of land-use layout is driven by a dynamic formula. The change process is based on an iterated function, which defines the increment g, depending on the initial patch amount:

$$g^t(n) = n + a_x \tag{1}$$

In this formula, t is the iteration times; n is the initial number of land-use patches (initial value = 0); a is a constant which is related to the value of x; x is the number of patches that have changed in one iteration. We set the patch that can only affect four neighbors by sides, so the value of x ranges from 1 to 4. The process is shown in Fig. 3:

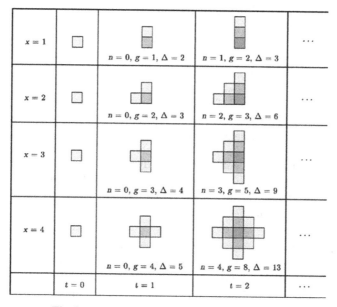

	$t = 0$	$t = 1$	$t = 2$	
$x = 1$		$n = 0, g = 1, \Delta = 2$	$n = 1, g = 2, \Delta = 3$	...
$x = 2$		$n = 0, g = 2, \Delta = 3$	$n = 2, g = 3, \Delta = 6$	...
$x = 3$		$n = 0, g = 3, \Delta = 4$	$n = 3, g = 5, \Delta = 9$	...
$x = 4$		$n = 0, g = 4, \Delta = 5$	$n = 4, g = 8, \Delta = 13$	...

Fig. 3. The general rules of the dynamics system

Due to the geometric features of the graphic figures (Fig. 3), the value of a is decided by the variation rules. Specific as follows:

- When $x = 1$ and $x = 4$, a takes x;
- When $x = 2$ and $x = 3$, a takes x in the first iteration, then a takes x $-$ 1 from the second iteration.

As a result of the patch randomly expanded in every iteration, the system will generate a more complicated graphical change. Because the function shares a similar feature with the dynamic formula of the fractal system, the results also hold the fractal features. Based on this dynamic formula, the sum value of certain patches Δ by time-varying function $\Delta(t)$ is:

$$\Delta(t) = \sum_{n=1}^{\infty} g^t(n) + 1 \tag{2}$$

In the model, the controlling over the number of various patches directly determines the results. With real-time apperception, the three types of agents can perceive and adjust the number of patches accordingly. This process is based on a judgment formula:

$$f(t) = \frac{n + \Delta(t)}{N} - R \tag{3}$$

which is:

$$f(t) = \frac{n + \left[\sum\limits_{n=1}^{\infty} g^t(n) + 1 \right]}{N} - R \qquad (4)$$

where N is the sum number of patches, R is the proportion of expected land-use, which is the upper development limit of the type of land use. Once R is lower than the expectation, the dynamic formula starts to work.

Behavioral Rule for Agents. According to the dynamics framework and the dynamics formula, all three agents have two behaviors: "moving" and "judging", and both behaviors alternate constantly. "Moving" refers to the random movement in the area, while "Judging" varies from each other as follows:

1. Developer: For the benefit of the industry development, the developer decides the allocation of industrial land according to the result of the land-use suitability evaluation. "Judging" behavior of developer means that the agent will check whether the land quality meets the requirement in the land-use suitability evaluation. If meets, the type of land use would alter to industrial land.
2. Residents: Residents will allocate public service facilities and green space based on accessibility. "Judging" behavior of residents means that the agent will check whether the proportion of different types of land use meets the control index of 15-min pedestrian-scale neighborhood. If not, the residents will change the layout of land use until all types of land use satisfy the requirement.
3. Government: The government will regulate the proportion of different types of land use. "Judging" behavior of the government means that if any kind of land use exceeds its expectation, the layout will be re-planned. The expectation includes the upper limitation of industrial development and the proportion of land use for housing, public service facilities, and green space. The government agent keeps dividing the site into circles of a radius of 800 m, counting the proportion of different land use in the circle, and adjusting the allocation until meeting the expectation.

Setting-Up Procedure. The user interface (Fig. 4) includes three parts: parameter input section, demonstration section, and observation section. The parameter input section is to input parameters to control the model operation. The demonstration section is to visualize the simulation of land use. The observation area is to observe quantitative relationships between different types of land use in real-time (Fig. 5).

1. Section 1 is the parameter input section. The parameters can be input include the scaling ratio, the number of agents, the speed, and the expected ratio of different kinds of land use. "Zoom" refers to the distance which is the side length of one patch, and the unit is meter. "num" represents the number of agents. "step" is the distance of every move, and the unit is meter. "ds", "gs", and "rs" refer to three different agents: developers, government, and residents. "rate" is the expected ratio of different types of land use, where "r" refers to industrial land in red color.

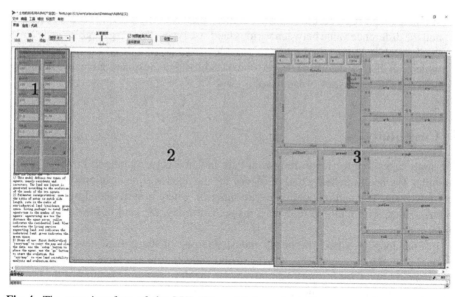

Fig. 4. The user interface of the LUL-IV model (overview) (1-parameter input section, 2-demonstration section, 3-observation section)

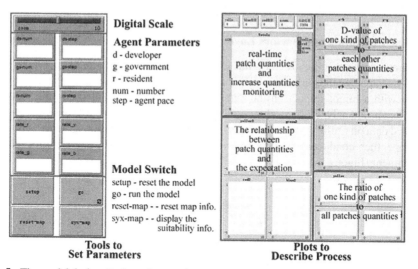

Fig. 5. The model design (Left section - tools to set parameters & Right section - plots to describe the process)

"y" refers to residential land in yellow color. "g" refers to green space in green color. "b" refers to public service facilities in blue color.

2. Section 3 is the observation section, which includes the increment graph of all types of land use ("Total"), the increment graph of different types of land use by proportion ("yellow", "green", "red", "blue"), the relationship between the number of

simulated land use and the expected value ("yellow2", "green2", "red2", "blue2") and the difference value between various land-use types ("rb", "rg", "gb", "yg", "yr", "yb", "rygb"). In the simulation process, the model keeps calculating and updating, which can display the fitting degree of expectation and the quantitative relationships between different types of land uses.

4   Case Study

The study takes Tianfo Health Industrial Park in Linyi City, Shandong Province as an example. The health industry and residence for the aged will be developed here in the next few years. The natural conditions are superior, where is surrounded by mountains on three sides, with multiple reservoirs inside. Currently, the land of the site is mostly for farming.

In Shandong province, the health industry is one of the critical developments in regional planning in recent years. Owing to abundant natural resources such as Yimeng Mountain in Linyi City, it is an essential place in the development planning of the health industry. Compared with other industries, the health industry is more sensitive to the land development, leading to the necessity of exploring the balanced land use through model analysis. The dynamics simulation results show the location and ratio of multiple land use, supporting land development decisions.

4.1   The Environmental Factors Layer of the Case

Data preprocessing is the crucial point in the environmental factors layer, which is to collect and analyze the current site information. The study is applied with an evaluation system with a minor impact on the natural environment. The impact on the industry development is analyzed by evaluating the cost of land development. Specifically, we collected data on traffic conditions, waterfront conditions, population, gradient, altitude, and slope, then conducted the GIS analysis based on both natural and socio-economic environment (Fig. 6).

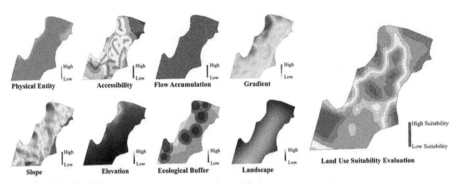

Fig. 6. Industrial land use suitability evaluation with weighted impact factors

4.2 Running LUL-IV Model

After constructing the environmental factors layer in the framework, we input the processed data to the model LUL-IV for simulation and set the parameters of the model evolution according to the actual needs of the industrial park project. Considering the evaluation of Industrial land-use suitability and the actual needs of the project, the expected ratio of industrial land is set as 30%. Figure 7 shows the first and second simulation process and outcomes with the same parameters setting:

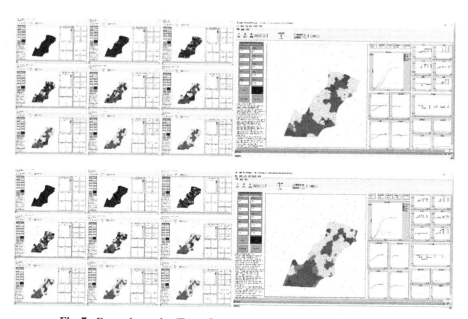

Fig. 7. Example results (Top – first outcome & Bottom - second outcome)

The observation section shows that the two simulation outcomes are consistent with the putative. Two simulations share similarities in trend and growth rate. Although there are differences between the visualization of the two simulations, several similar trends still exist. Not only in these two simulations, but we also find that the results of every simulation have some features in common, these commonalities include:

- At the beginning of the simulation, the growth rate of industrial land is rather slow compared with other sorts of land use. We suppose that developers have to spend time selecting and deciding the land to be developed in the early stage;
- During the simulation process, residents will allocate excessive green space and public service facilities in the 15-min neighborhood, which will be regulated to a reasonable value range by the government;
- The growth of residential land is more stable than other types of land use.

Theoretically, the outcome of each simulation is the optimal layout for the situation. And then, the state of the system can be observed from the simulation results, which were obtained through a batch of simulations. Through the superposition and pixel clustering of 500 times simulation results, it indicates the spatial distribution of housing, industry, infrastructure, and green space.

Fig. 8. Model outcomes. (Left - aggregated outcome of 500-time simulations & Right - clustered outcome)

4.3 Simulation Results

Through the dynamics simulation and multi-agent model LUL-IV, the planning of the Tianfo Industrial Park is supported by a general land-use layout that meets the need of residents, developers, and government. Specifically, the following features can be found in the aggregated land-use layout (generated by aggregating 500 simulations) (Fig. 8):

- The boundaries of different types of land use are presented in an organic form, and they are interspersed with each other;
- Industrial land (in orange color) is agglomerated, with two cores distributes in the north and south;
- The residential land is mainly in area distribution;
- The green space and public service facilities are dispersed in points;
- The distribution of land for housing, green space, and public service facilities tends to be even distribution.

Although the interactions among the agents are complex and random, it still can be explained by the mechanism of the dynamics system. Firstly, the gaming where residents and developers are under the regulation of government leads to an incomplete match between the spatial distribution of industrial land and the land-use suitability evaluation. Secondly, it also results in the residential construction occupying a part of high-quality areas. Moreover, the spatial distribution of different types of land use is closely related to the shape of the site, especially meeting the requirement of the residential land use layout, the 15-min neighborhoods.

In this case, the results of the simulation are compatible with the original land development intention. Because the model generation is anchored on the evaluation of land use, the results are also compatible with the carrying capacity of the land. The actual land-use planning of the industrial park took the result of the study as a reference: planners amended the planning and re-decided the land-use spatial structure in the Tianfo Industrial Park inspired by this study to some extent. The industrial land of the park gathers at two points, developing the two spatial cores and axis of the park. The spatial distribution of housing, green space, and public service facilities promotes the decision-making of detailed planning. The traditional planning methods focus on the evaluation of resources. While in this case, not only does it consider the results of traditional land-use evaluation, but also it simulates the process of land development. Hence, the final results comparatively better respond to the top-down planning of the government and the bottom-up needs of residents and developers.

The research takes a system dynamics framework as its working basis. Compared to previous planning decisions from experiences based on bounded rationality assumptions, the simulation method is more logical for applying the action rules into planning. We also find that the behaviors of agents in the model are close to the decision-making process in real urban development planning. Broadly, the significance of the simulation is that the application of the model can enable planners to understand the needs of people and cities in land development, thus balance these needs to reach an optimal and sustainable plan.

5   Conclusions

Industrial development contributes to urban development [10]. However, in practice, the effectiveness of the industrial land development is not entirely satisfied, sometimes even conflicts with residents' needs. In order to understand this complex process, the study constructed a system dynamics framework. Under this framework, we analyzed the relationship and interactions between developers, residents, and the government as theoretical support of the simulation.

This research is trying to help the planners to balance all participants' needs and benefits during the urban development, especially in the process of land development. The case of Tianfo Industrial Park illustrates the modeling process and provides a reference for the land use planning. By this research framework and multi-agents model LUL-IV, planners and stakeholders can directly observe the changes and distribution of different types of land use visually and quantitively. The framework and model can also be adapted to the land use planning of other industrial parks. Furthermore, if the system can be simulated in more cases, by comparing simulation results of different cases, it is possible to discover the fundamental laws in the land use planning and improve the accuracy of the model.

However, the evolution of urban land use is an extremely complicated process, and we have to acknowledge the limitation of the model, such as residents of different ages and gender may have different needs for the living environment. To achieve more precise simulation, future research direction should be placed on more subdivided types of agents.

References

1. Benenson, I., Torrens, P.M.: Geographic automata systems: a new paradigm for integrating GIS and geographic simulation. Presented at the AGILE 2003: 6th AGILE Conference on Geographic Information Science (2003)
2. Li, X., Ye, J.: Cellular automata for simulating complex land use systems using neural networks. Geogr. Res. **01**, 19–27 (2005)
3. Liu, H., Silva, E.: Simulating the dynamics between the development of creative industries and urban spatial structure: an agent-based model. In: Geertman, S., Toppen, F., Stillwell, J. (eds.) Planning Support Systems for Sustainable Urban Development, pp. 51–72. Springer, Heidelberg (2013). https://doi.org/10.1007/978-3-642-37533-0_4
4. Liang, Y., Li, W., Jiang, C., Liu, L.: The research frame of spatial decision-making process of enterprise migration based modelling. Econ. Geogr. **34**(04), 112–118 (2014)
5. Rai, V., Robinson, S.A.: Agent-based modeling of energy technology adoption: empirical integration of social, behavioral, economic, and environmental factors. Environ. Model. Softw. **70**, 163–177 (2015)
6. Wu, Z., Gan, W.: Urban intelligent planning technology practice in transitional period. Urbanism Archit. (1), 26–29 (2018)
7. Wang, W., Yang, S., Hu, F., Han, Z., Carlo, J.: An agent-based modeling for housing prices with bounded rationality. Presented at the Journal of Physics: Conference Series (2018)
8. Liu, H.: Research on Urban Land Expansion and Planning Scenarios Simulation (2009)
9. Geertman, S., Toppen, F., Stillwell, J. (eds.): Planning Support Systems for Sustainable Urban Development. Lecture Notes in Geoinformation and Cartography. Springer, Heidelberg (2013). https://doi.org/10.1007/978-3-642-37533-0
10. Miao, Z., Chen, Y., Zeng, X.: Urban spatial layout intelligent model research based on multi-agent system. J. Inf. Comput. Sci. **10**(14), 4627–4637 (2013)
11. Li, M.: The planning strategies of a 15-minute community life circle based on behaviors of residents. Urban Plann. Forum **1**, 111–118 (2017)
12. Xie, Z.: Study on the Urban Spatial Growth Model Based on Multi-Agent System. Nanjing University (2015)
13. Xie, Z., Li, F., Li, M., Chen, Z., Zhou, C.: Multi-agent simulation model of urban spatial growth under government planning constraints. Geogr. Geo-Inf. Sci. **31**(02), 60–64+69+127 (2015)
14. Shan, Y., Zhu, X.: Scenario analysis of urban residential land use utility based on multi-agents' spatial decision. J. Nat. Resour. **26**(11), 1832–1841 (2011)
15. Shan, Y.Z., Zhu, X.: Multi-agents model for simulation of urban residential space evolution. Prog. Geogr. **30**(08), 956–966 (2011)
16. Ji, M., Monticino, M., Acevedo, M.: Multi-agent based modeling of land-use decision making process in a democratic setting. Geogr. Res. **28**(01), 85–96 (2009)

Attention U-Net for Road Extraction in Remote Sensing Images

Minyu Tao[1], Zhiming Ding[1,3(✉)], and Yang Cao[2]

[1] Beijing University of Technology, Beijing 100124, China
`taomy@emails.bjut.edu.cn`
[2] Beijing Wuzi University, Beijing 101149, China
`caoyangcwz@126.com`
[3] Institute of Software, Chinese Academy of Sciences, Beijing 100190, China
`zhiming@iscas.ac.cn`

Abstract. The reliable road network plays a vital role in many applications. Owing to the development of remote sensing technology and the success of deep learning in computer vision, automatic road extraction from remote sensing images is a research hotspot in recent years. However, due to the complicated image background and special road structure, the results of automatic road extraction are still far from perfect. In this paper, we propose a road segmentation network that is designed based on improved U-Net, which contains an encoder and a decoder. First, the recurrent criss-cross attention module (CCA) is introduced into the encoder to obtain long-range contextual dependencies with a relatively small number of computations and parameters, which results in better understanding and expression of image information. Second, we propose the attention-based multi-scale feature fusion module (AMS) to resolve the problem of different shapes and widths of the roads, which is placed between the encoder and decoder and uses attention mechanisms to guide multi-scale information fusion. Experimental on the Massachusetts Roads Dataset show that the proposed method achieves better performance in road extraction than other methods in terms of precision, recall, F1-score, and accuracy.

Keywords: Remote sensing · Attention mechanisms · Road extraction

1 Introduction

The road network as a kind of infrastructure has been widely used in both military and civil. In addition, with the rapid increase of remote sensing image data, a large amount of geographic information is obtained from remote sensing images, which is beneficial to road network updating, urban planning, traffic navigation, disaster relief, and so on [1]. Thus, there is a high demand for reliable road extraction methods based on remote sensing images.

This paper defines the road extraction problem as the image segmentation problem, which means that we need to determine whether each pixel belongs to the road pixels. Nevertheless, compared with the common target segmentation, road extraction based on remote sensing images is more challenging. Firstly, the background of remote sensing images is complicated. It is susceptible to interference from other objects. Secondly, the road structure is diversified, and roads vary in width and shape. Thirdly, images are taken from high altitude, causing the problem of occlusion. These factors make the results of road extraction less reliable.

To solve the above challenges, researches on the use of remote sensing images to extract roads are extensive. The traditional method mainly used the characteristics of texture, spectrum, geometric for road extraction [2–4], which perform poorly in generalization at different data sets and tend to make a lot of noise in the results. Recently, deep learning has made significant progress in automatic road extraction. Many commonly used networks have achieved excellent results in road extraction, such as FCN [5], U-Net [6], DenseNet [7], and so on. However, these models also have their limits, which is easy to ignore the relationship between pixels and can only capture short-range contextual dependencies because of the structure. Thus, these models usually further optimize road extraction results with post-processing methods. Moreover, because of the particularity of road structure and the complexity of background, local contextual information is insufficient to deal with road extraction problems, which leads to poor performance in road extraction. Therefore, it is necessary to propose a novel network to obtain global contextual dependencies in road extraction. Considering that attention mechanisms achieve excellent performance in image segmentation, such as the Nonlocal module [14], PSANet [15], DANet [16]. Through attention mechanisms, we can aggregate global contextual information from all pixels to better understand and express the image information. In addition, attention mechanisms can guide the network to focus on the most relevant features and suppress irrelevant features. Motivated by this, we propose our new network by integrating the attention modules.

Our goal is to extract reliable and detailed road maps. In this paper, we propose a new road segmentation network based on the U-Net with integrated attention mechanisms. First, roads usually occupy the whole image because of the slender character, and buildings and trees tend to block roads. Therefore, obtaining the long-range dependencies is necessary by the road extraction network. Inspired by recent work on the recurrent criss-cross attention module (CCA) [18], we introduce the CCA in the encoder of U-net, which can capture global contextual information and reduce the distance between each pair-pixel. Second, roads vary in shape and width, which also affects the results of the road extraction. It is necessary to integrate contextual information at different scales. To solve the problem, we propose the attention-based multi-scale feature fusion module (AMS) to capture multi-scale information, which can adaptively weigh the importance of information at different scales by integrating attention mechanisms. The model is placed between the encoder and the decoder. Additionally, combining attention mechanisms in the proposed method makes the network more focused on learning important information and suppress irrelevant information. Experiments showed that our approach is superior to others.

The main contributions of this paper are as follows:

(1) For the first time, the recurrent criss-cross attention module (CCA) is introduced into the road segmentation network to obtain the long-range contextual dependencies from all pixels and effectively solve the problem of road information loss.

Secondly, compared with other self-attention modules, the number of parameters and calculations are reduced through recursive criss-cross operations.

(2) The attention-based multi-scale feature fusion module (AMS) is proposed to selectively integrate important multi-scale context and effectively solve the problem of different road shapes and widths.

(3) According to the complexity of remote sensing images and the special structure of roads, improved U-Net based on attention mechanisms. First, the integrated recurrent criss-cross attention module during encoder can access the global contextual dependencies. Second, the attention-based multi-scale feature fusion module is placed between the encoder and decoder to solve the multi-scale problem. Based on the above improvements, our method pays more attention to learn road characteristics and effectively improves the accuracy of the road extraction.

The structure of this paper is as follows. Section 2 summarizes the related work. Section 3 introduces the proposed overall architecture and illustrates some of the theoretical knowledge. Section 4 describes and analyzes the experiment results of road segmentation. Finally, Sect. 5 concludes this work.

2   Related Works

In this section, we specifically describe the current mainstream methods of road extraction, which mainly divided into traditional methods and deep learning methods.

Traditional methods mainly include spectral analysis, edge detection, and threshold segmentation, and so on, for example, Shi et al. [2] used adaptive neighborhoods to classify spectral features and spatial features, which can segment the images into roads and non-roads, but the method required training of SVM for each input image and was not suitable for complex prototype intersections. Huilin et al. [3] proposed to generate binary images with the target and background based on the gray features, but it was not fast enough to identify the road and difficult to distinguish the irrelevant features. Gaetano et al. [4] first proposed using the Canny operator for edge extraction, then obtained the final road maps by graph cutting theory. Most of these traditional methods are generally time-consuming and have weak generalization ability, which usually requires human intervention and highly depends on threshold parameters.

With the rapid development of artificial intelligence, it is popular to use computer vision algorithms to realize automatic road extraction. Based on the FCN network architecture, Wei et al. [8] added the road structure information into the loss function to improve the road segmentation results. Kestur et al. [9] proposed the UFCN model to realize road extraction in low-altitude remote sensing images. The model consists of a set of convolution stacks and corresponding mirror deconvolution stacks and uses the shortcut connection to save local features. Zhang et al. [10] proposed a road extraction network combining the advantages of ResNet [11] and U-Net [6], which was composed of the residual module and simplified the training of the deep network. Zhao et al. [12] proposed the Road-RCF model for road extraction from HSRRS images, where RCF as the post-processing method was used to make better results in road extraction. Jiang et al. [13] proposed DenseUNet to realize road extraction, which could aggregate features at different scales by dense connection units and skips connections. The above road extraction methods based on FCN made significant progress by increasing the number of convolution layers or combining multiple networks, which usually require plenty of computations for getting bigger receptive field and richer contextual information. Moreover, these methods can not aggregate long-range contextual dependencies from all pixels, resulting in the loss of road information. So these methods cannot meet our requirements in road extraction from remote sensing images.

Recently, attention mechanisms have been widely used in various deep learning tasks such as NLP, image recognition, and so on. SENet greatly improved the expression ability of the network by establishing channel attention mechanisms to allocate weights to each channel [17]. Wang et al. [14] proposed a non-local module with intensive context information aggregation, which generates a huge attention map by calculating the relation matrix between features. Huang et al. [18] proposed the recurrent criss-cross attention model to obtain long-range dependencies from all pixels more efficiently and effectively. Therefore, we propose the CCA module and ASM module based on attention mechanisms, which can effectively solve the problems in road extraction and optimize the results of road extraction.

3   Proposed Method

3.1   Overall Architecture

In this section, we will introduce the proposed road extraction network in detail. As shown in Fig. 1, the proposed method is encoder-decoder architecture. We choose U-net as the backbone because of its ability to recover fine detail by fusing low-level spatial information and high-level semantic information. In the encoding phase, we decide ResNet34 as the encoder to extract features, which is good at solving the problem of information loss through residual learning. Besides, the recurrent criss-cross attention module (CCA) is integrated into the encoder to take full advantage of global contextual information for road extraction. The attention-based multi-scale feature fusion module (AMS) is added between the encoder and decoder to use attention mechanisms guiding multi-scale feature integration, which can effectively solve the multi-scale problem in the road extraction task. The decoder fuses the features generated by the encoder at different stages, and low-level spatial information from encoder can supplement the high-level semantic information for more detailed results. Finally, road masks of the same size as the original images are generated by up-sampling in the decoder.

3.2   Recurrent Criss-Cross Attention Module

Motivated by the performance of self-attention in image segmentation, the self-attention module can capture long-range dependencies and focus on learning essential features by assigning weights to each feature [19]. Considering that roads occupy the whole image and roads have the characteristic of being long and thin, the integration of attention mechanisms in the network is beneficial to road extraction. However, these self-attention modules generate huge attention maps, which cause a considerable amount of calculations and parameters, and the complexity of these modules in space and time are

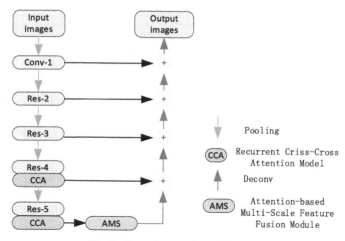

Fig. 1. Overall network architecture.

both $O(w \times h) \times (w \times h)$. The recurrent criss-cross attention model can also capture contextual information from all pixels, whose complexity in space and time are $O(w \times h) \times (w + h - 1)$ [18]. Thus, in the proposed framework, we introduce the recurrent criss-cross attention module (CCA) into the encoder to capture intensive long-range contextual dependencies with lower computational and storage costs.

Figure 2 shows the architecture of CCA. The feature map X is the input of the module with the size of $c1 \times w \times h$, where $c1$ represents the number of channels, w represents the width, and h represents the height. Two feature maps Q and K are generated respectively by applying the convolution layer with a kernel size of 1×1, with dimensions of $c2 \times w \times h$, where $c2$ is smaller than $c1$. Then Q and K are used to generate the attention map A that represents the importance of each feature through Affinity operation to get horizontal and vertical information, with the size of $(w + h - 1) \times w \times h$. Similarly, X goes through another convolution layer of 1×1 to generate V, with the size of $c1 \times w \times h$. Finally, X' is obtained through the aggregation operation of feature maps V and A. However, X' is only able to get information in the horizontal and vertical direction, which does not satisfy the requirement of collecting the relationships between each pixel-pair. To capture richer contextual information, X' is fed into the module to generate a new feature map X", which can gather dense contextual information from all pixels. Parameters are shared in two recurrent criss-cross operations to reduce extra parameters. Considering that adding the recurrent criss-cross attention module will increase the computation, try to put it in the back of the encoder. The integration of CCA in the encoder enables the network to obtain sufficient contextual information to avoid the loss of road information.

3.3 Attention-Based Multi-scale Feature Fusion Module

Because the roads vary in shape and width, it is of great importance to obtain multi-scale features of the road. The ASPP module uses multiple dilation Convolution with different dilation rates to collect contextual information at multiple scales [20]. However,

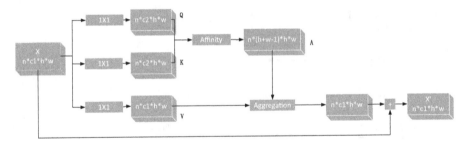

Fig. 2. Criss-cross attention module.

the ASPP module does not consider the importance of contextual information from different scales and lacks the flexibility to model information at different scales.

To solve the above problem, we propose the attention-based multi-scale feature fusion module (AMS), which uses the attention mechanisms to guide multi-scale information fusion and avoid information redundancy. Figure 3 shows the structure of the AMS module. Firstly, the multi-scale contextual information is obtained through four 3×3 convolutions with different dilation rates. Secondly, the global contextual information of the feature map x is used to weigh multi-scale information at different scales. Finally, the module uses the weights to guide multi-scale features fusion, which allows the module to select more relevant features and suppress irrelevant features. With the attention-based multi-scale feature fusion module, we can solve the multi-scale problem in road extraction and improve the performance of the road extraction.

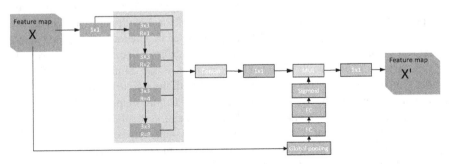

Fig. 3. Attention-based multi-scale feature fusion module.

We will describe the attention-based multi-scale feature fusion module in detail below. First, the feature map X is fed into a 1×1 convolution layer as a bottleneck to reduce the number of channels. Second, the feature map passes through four convolutional layers of 3×3 with the dilation rate of $1, 2, 4, 8$ successively to obtain multi-scale contextual information at different scales. Then we concatenated the collected multi-scale contextual information and fed it into another 1×1 convolution to generate a new feature map y, which collections multi-scale contextual information, as shown in Eq. (1).

$$y = (\sum_i x_i w_i) v \tag{1}$$

Next, the global context of feature map X is used to generate the channel description G through the global average pooling. Vector G is generated by shrinking X. Formula (2) represents the primary process, where c represents the number of channels, h and w represent the height and width, respectively.

$$G_c = \frac{1}{h \times w} \sum_{i=1}^{h} \sum_{j=1}^{w} x_c(i,j) \qquad (2)$$

The feature map G generated by global average pooling is fed into the two FC layers to adjust the weights adaptively, which makes the weights learned more generalized. The first FC layer reduces the number of channels by compression ratio $k = 8$, then followed by the ReLU function. The second FC layer makes the output dimensions match the channels number of y, followed by sigmoid activation, mainly to capture the dependencies between channel information. The weights of the final output represents the importance of each channel. The resulting channel attention map A is obtained by formula (3), where φ represents the sigmoid function, and δ represents the ReLU function.

$$A = \varphi(w_2 \delta(w_1 G)) \qquad (3)$$

After the channel attention map A is obtained, multiply the corresponding channel between feature map y and channel attention map A to guide the multi-scale fusion. Use the weight to select the vital multi-scale features and focus on the important information, following a 1×1 convolution to enhance information interaction between channels. The final output of this module is shown in Eq. (4).

$$x' = (y \cdot A)w \qquad (4)$$

4   Experiment

4.1   Datasets

In this work, we employ Massachusetts Roads Dataset as experimental data that was built by Mihn [21], including 1108 training images, 14 validation images, and 49 test images. The size of each image is 1500×1500, with a resolution of 1 m/pixel. Besides, the data set belongs to the aerial remote sensing image data set, covering an area of more than 2634 km^2.

4.2   Experimental Details

We used ImageNet pre-trained ResNet34 to be the encoder of the network and remove the last full connection layer. The CCA and AMS are added to the U-Net to improve the accuracy of road extraction. To avoid the problem of over-fitting, we did data augmentation, including image shifting, horizontal flip, vertical flip, ambitious color jittering, and so on during training. The network is implemented based on the PyTorch framework.

Given the imbalance of positive and negative samples in the road segmentation task, the proportion of road pixels is much smaller than the background pixels, which causes a large amount of information to be lost during iteration. Therefore, we chose dice coefficient loss and binary cross entropy loss as the loss function and used Adam optimizer to update the parameters adaptively. The batch size was fixed as 16. The learning rate was initially set 2e−4 and divided by five while observing the training loss decreasing slowly until the model convergence. The model was trained simultaneously on four 2080ti GPU.

As the test images of the Massachusetts Roads Dataset are small, we also did data augmentation on test images to improve the robustness of the prediction, including images horizontal flip, vertical flip, and diagonal flip.

4.3 Evaluation Criteria

In this work, accuracy, precision, recall, and F1-score were used to evaluate the road segmentation performance. Among them, accuracy is the percentage of the samples that are predicted correctly among all samples. Precision is the ratio of the actual road samples among all samples predicted to be roads. The recall represents the percentage of samples that are actually predicted to be positive among all road pixels. Considering precision and recall are contradictory, F1-score hit the right compromise between precision and recall to achieve the maximum of both.

$$accuracy = \frac{TP + TN}{TP + FN + FP + TN} \tag{5}$$

$$precision = \frac{TP}{TP + FP} \tag{6}$$

$$recall = \frac{TP}{TP + FN} \tag{7}$$

$$F1 - score = \frac{2 \times precision \times recall}{precision + recall} \tag{8}$$

TP, FP, TN, and FN represent the true positive, false positive, true negative, and false negative pixels of the predicted image, respectively.

4.4 Results and Discussion

In this section, we demonstrate the effectiveness of our proposed network integrated with CCA and AMS in road extraction. First of all, we evaluate and compare the proposed network with other state-of-the-art methods using the above evaluation criteria. Secondly, we also evaluate the role of recurrent criss-cross attention module and attention-based multi-scale feature fusion module in road extraction.

In order to prove the effectiveness of the proposed model, our model is compared with some state-of-the-art methods, including DeepLabV3+ [22], U-Net and D-Linknet [23]. The experimental constraints are the same.

Table 1. Evaluation results of different models on the Massachusetts Roads Dataset.

Model	Precision (%)	Recall (%)	F1-score (%)	Accuracy (%)
DeepLabV3+	71.802	72.657	72.630	97.120
U-Net	69.712	79.630	74.176	97.139
D-LinkNet	69.975	81.159	74.917	97.219
Proposed	71.822	80.740	75.735	97.378

Table 1 shows the road extraction results of different methods on the Massachusetts Roads Dataset that are evaluated by precision, recall, F1-score, and accuracy, respectively. As can be seen from the evaluation results of each model, our method achieves a precision of 71.822%, recall of 80.740%, F1-score of 75.735%, and accuracy of 97.378%. Notably, our model achieves the highest precision, F1-score, and accuracy, which means that our model is better able to avoid the influence of the background factors similar to the roads and distinguish road pixels from Non-road pixels more accurately. Moreover, our model is second only to D-LinkNet in recall rate, which is caused by a contradiction between precision and recall. We balance these two criteria by F1-score. In terms of F1-score, our model was 3.105%, 1.559%, and 0.818% higher than DeepLabV3+, U-Net, and D-LinkNet, respectively. Compare to other models, our model performs outstanding and can better identify road pixels in the task of road extraction.

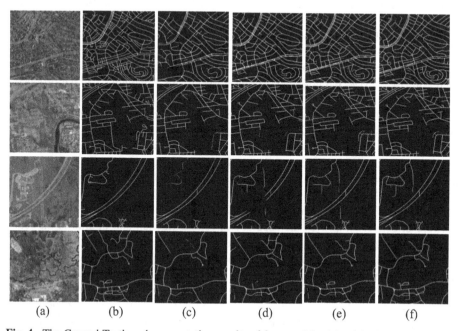

(a)	(b)	(c)	(d)	(e)	(f)

Fig. 4. The Ground Truth and segmentation results of four models. (a) original images. (b) the ground truth. (c) DeeplabV3+. (d) U-Net. (e) D-LinkNet. (f) proposed.

The visual performance of road extraction in the Massachusetts Roads Dataset by different models is shown in Fig. 4. As we can see from these examples, the results of other methods are more influenced by the background. For example, some objects that have similar shapes or same spectrum with the road are easily mistaken as the road, and there is the phenomenon of road information loss, while road extracted by our method is closer to the ground truth, including more detail for both sparse roads and dense roads. By comparing the visual results, our approach achieved relatively satisfactory performance in terms of continuity and completeness. In addition, our method can also identify some smaller roads, which suggests that our method is superior to other methods in different scale road extraction.

Table 2. Evaluation results of adding the different modules.

CCA	AMS	Precision (%)	Recall (%)	F1-score (%)	Accuracy (%)
×	×	69.727	81.378	74.832	97.203
√	×	71.626	79.453	74.995	97.328
×	√	71.958	79.434	75.186	97.347
√	√	71.822	80.740	75.735	97.378

We also demonstrate the effectiveness of the recurrent criss-cross attention module and attention-based multi-scale feature fusion module. Table 2 illustrates the evaluation results of integrating different modules in road extraction on the Massachusetts Roads Dataset. In evidence, the model could only achieve 69.727% precision, 81.378% recall, 74.832% F1-score, and 97.203% accuracy without integrating CCA and AMS. The model improved significantly by adding ASM in precision, F1-score, and accuracy, increasing by 2.231%, 0.354%, and 0.144%, respectively. Although recall has dropped, precision has improved, we usually want a balance between precision and recall, which can be represented by F1-score. The increase in the F1 score indicates that our network performs better than others. As can be seen from Table 2, both F1-score and accuracy increased significantly after integrating ASM. Besides, the model has been further improved after adding the recurrent CCA. By combining these two modules, our model achieves the best performance in road extraction. The above experiments prove that the recurrent CCA and AMS play an essential role in the task of road segmentation.

In summary, it can be concluded that the introduction of CCA and AMS into the network effectively improves the accuracy of road extraction and optimizes the road segmentation details. The application of the recurrent CCA can obtain long-range dependencies from all pixels, and we can better understand and express the image information, and solve the problem of road discontinuity and avoid interference by similar objects effectively. AMS integration is also crucial to solving the multi-scale problem, which can focus on the most relevant information and inhibit irrelevant information while fusing multi-scale information and avoid the redundant use of similar information. Compared with other methods, our model can better prevent the interference of other objects and reduce the noise from road segmentation results effectively. Experiments show that the improvements are suitable for road extraction tasks and have made significant progress.

5 Conclusions

In this paper, a new network with better understanding and expression ability is proposed to implement road extraction based on remote sensing images. Compared with the current popular segmentation network, our method has better performance in road extraction. The significant contribution of this paper is to improve the U-Net by combining the attention mechanism to make it have better performance in road extraction. The encoder-decoder structure performs well in extracting road details, where pre-trained ResNet34 acts as the encoder to solve the degradation problem during training. The recurrent criss-cross attention module is introduced into the encoder to obtain intensive global contextual dependencies. The attention-based multi-scale feature fusion module is located between the encoder and decoder to solve the problem of different road shapes and widths. The model with dilation convolution can effectively capture multi-scale features and use attention mechanisms to guide multi-scale information fusion. Experimental results on the Massachusetts Roads Dataset demonstrated that our method outperformed the compared methods in terms of precision, recall, F1-score, and accuracy.

Considering that the proposed method needs to be trained in different environments to improve its robustness, the network will have better expression ability and generalization ability by increasing the training set. Therefore, we will try larger data sets in the future to further improve the performance of the network.

References

1. Jiang, Y.: Research on road extraction of remote sensing image based on convolutional neural network. EURASIP J. Image Video Process. **2019**, 31 (2019). https://doi.org/10.1186/s13 640-019-0426-7
2. Shi, W., Miao, Z., Debayle, J.: An integrated method for urban main-road centerline extraction from optical remotely sensed imagery. IEEE Trans. Geosci. Remote Sens. **52**(6), 3359–3372 (2014)
3. Mu, H., Zhang, Y., Li, H., Guo, Y., Zhuang, Y.: Road extraction base on Zernike algorithm on SAR image, pp. 1274–1277 (2016). https://doi.org/10.1109/IGARSS.2016.7729323
4. Gaetano, R., Zerubia, J., Scarpa, G., et al.: Morphological road segmentation in urban areas from high resolution satellite images. In: Proceedings of International Conference on Digital Signal Processing, pp. 1–8 (2011)
5. Long, J., Shelhamer, E., Darrell, T.: Fully convolutional networks for semantic segmentation. IEEE Trans. Pattern Anal. Mach. Intell. **39**(4), 640–651 (2014)
6. Ronneberger, O., Fischer, P., Brox, T.: U-Net: convolutional networks for biomedical image segmentation. In: Navab, N., Hornegger, J., Wells, W., Frangi, A. (eds.) Medical Image Computing and Computer-Assisted Intervention – MICCAI 2015. MICCAI 2015. Lecture Notes in Computer Science, vol. 9351, pp. 234–241. Springer, Cham (2015). https://doi.org/10.1007/978-3-319-24574-4_28
7. Huang, G., Liu, Z., van der Maaten, L., Weinberger, K.Q.: Densely connected convolutional networks (2017)

8. Wei, Y., Wang, Z., Xu, M.: Road structure refined CNN for road extraction in aerial image. IEEE Geosci. Remote Sens. Lett. **14**(5), 709–713 (2017)
9. Kestur, R., Farooq, S., Abdal, R., Mehraj, E., Narasipura, O.S., Mudigere, M.: UFCN: a fully convolutional neural network for road extraction in RGB imagery acquired by remote sensing from an unmanned aerial vehicle. J. Appl. Remote Sens. **12**, 016020 (2018)
10. Zhang, Z., Liu, Q., Wang, Y.: Road Extraction by Deep Residual U-Net. IEEE Geosci. Remote Sens. Lett. **15**, 749–753 (2018)
11. He, K., Zhang, X., Ren, S., et al.: Deep residual learning for image recognition. In: 2016 IEEE Conference on Computer Vision and Pattern Recognition (CVPR). IEEE Computer Society (2016)
12. Hong, Z., Ming, D., Zhou, K., Guo, Y., Lu, T.: Road extraction from a high spatial resolution remote sensing image based on richer convolutional features. IEEE Access **6**, 46988–47000 (2018)
13. Xin, J., Zhang, X., Zhang, Z., Fang, W.: Road Extraction of High-Resolution Remote Sensing Images Derived from DenseUNet. Remote Sens. **11**(21), 2499 (2019)
14. Wang, X., Girshick, R.B., Gupta, A., He, K.: Non-local neural networks. In: CVPR, pp. 7794–7803 (2018)
15. Zhao, H., et al.: PSANet: point-wise spatial attention network for scene parsing. In: ECCV, vol. 9, pp. 270–286 (2018)
16. Fu, J., Liu, J., Tian, H., Li, Y., Bao, Y., Fang, Z.: Dual attention network for scene segmentation (2019)
17. Hu, J., Shen, L., Sun, G.: Squeeze-and-excitation networks. In: The IEEE Conference on Computer Vision and Pattern Recognition (CVPR), pp. 7132–7141 (2018)
18. Huang, Z., Wang, X., Huang, L., et al.: CCNet: criss-cross attention for semantic segmentation (2018)
19. Wang, Y., Seo, J., Jeon, T.: NL-LinkNet: toward lighter but more accurate road extraction with non-local operations. IEEE Geosci. Remote Sens. Lett., 1–5 (2019)
20. Chen, L.C., Papandreou, G., Schroff, F., et al.: Rethinking atrous convolution for semantic image segmentation (2017)
21. Mnih, V., Hinton, G.E.: Learning to detect roads in high-resolution aerial images. In: Daniilidis, K., Maragos P., Paragios, N. (eds.) Computer Vision – ECCV 2010. ECCV 2010. Lecture Notes in Computer Science, vol. 6316, pp. 210–223. Springer, Heidelberg (2010). https://doi.org/10.1007/978-3-642-15567-3_16
22. Chen, L.-C., Zhu, Y., Papandreou, G., Schroff, F., Adam, H.: Encoder-decoder with atrous separable convolution for semantic image segmentation, p. 4. arXiv:1802.02611 (2018)
23. Zhou, L., Zhang, C., Wu, M.: D-LinkNet: LinkNet with pretrained encoder and dilated convolution for high resolution satellite imagery road extraction. In: 2018 IEEE/CVF Conference on Computer Vision and Pattern Recognition Workshops (CVPRW). IEEE (2018)

College Students' Portrait Technology Based on Hybrid Neural Network

Zhiming Ding$^{(\boxtimes)}$ and Xuyang Li

Beijing University of Technology, Beijing 100022, China
zmding@bjut.edu.cn, lixuyang@emails.bjut.edu.cn

Abstract. Students have produced a large number of data in the teaching life of colleges and universities. At present, the development trend of university data is to gradually form a high-dimensional data storage system composed of student status information, educational administration information, behavior information, etc. It is of great significance to make use of the existing data of students in Colleges and universities to carry out deep-seated and personalized data mining for college education decision-making, implementation of education and teaching programs, and evaluation of education and teaching. Student portrait is the extension of user portrait in the application of education data mining. According to the data of students' behavior in school, a labeled student model is abstracted. To address above problems, a hybrid neural network model is designed and implemented to mine the data of college students and build their portraits, so as to help students' academic development and improve the quality of college teaching. In this paper, experiments are carried out on real datasets (the basic data of a college's students in Beijing and the behavior data in the second half of 2018–2019 academic year). The results show that the hybrid neural network model is effective.

Keywords: Big data · Data mining · Higher education ·
Convolutional neural network · Feedforward neural network

1 Introduction

In the past few years, revolutionary changes have occurred in the field of education and information. Online learning systems, smartphone applications and social networks have provided a large number of applications and data for educational data mining (EDM) research. Massive open online courses (MOOCs) became a new type of teaching model in the recent 2 years. It's obviously that EDM is in the era of "big data", which tells that EDM research will be developed rapidly [1].

As far as the current situation is concerned, it is very meaningful to carry out a deep-level and personalized data mining for the existing student data of

Supported by the National Key R&D Program of China under grant number 2017YFC0803300.

X. Meng et al. (Eds.): SpatialDI 2020, LNCS 12567, pp. 165–183, 2021.
https://doi.org/10.1007/978-3-030-69873-7_12

colleges and universities. Student portraits are an extension of user portraits in educational data mining applications, which makes an abstract label student model according to the data of students' behavior in school.

The establishment of a student portrait system in colleges and universities has the following significance: the adjustment of educational decision-making has avoided problems such as partial models and inaccurate data due to the lack of analysis tools; the correct implementation of education and teaching programs has avoided large differences due to students In the management of students, attention is paid to the classification and partitioning and dynamic adjustment, which affects the teaching effect; to achieve efficient management of students, real-time understanding of students' thinking dynamics, and accurate management of students.

At present, many universities collect and display the relevant data generated by students' work, instead of research on data mining. In terms of student portraits, their research is more about generating data for students The statistics on the number and frequency of information. The portrait tags are more derived from relevant higher education researchers based on experience and are not comprehensively representative.

Therefore, this paper aims at the basis of hybrid neural networks, based on some basic information (birthplace, ethnicity, gender, etc.) currently known to undergraduates in a certain college and university, the educational administration information data generated by students during the semester, and the behavior information data generated by campus card and gateway account on the basis of hybrid neural network for students to help students academic development and to improve the teaching quality of colleges and universities.

2   Related Work

During the preliminary investigation, we found that the current research on EDM is not enough in universities, and most of them stay on the statistics, collation and summary of student data, or use relatively simple machine learning methods, such as decision tree classifiers and naive Bayes classifiers [2,3], it is impossible to predict and analyze the potential problems or tendencies of students. In horizontal comparisons in other industries, we found that a large number of deep learning algorithms have been applied to user data mining in industries such as e-commerce, socialization, and finance [4–6]. And text classification, we learned from this and optimized it, and proposed a portrait technology for college students based on hybrid neural networks.

2.1   Portrait Technology

The concept of "user portrait" was first proposed by Alan Cooper, the father of interaction design [7]. It refers to an abstract labelled user model. This model is based on basic user information, social information, preference information and behavior. The information is summarized. In the process of forming user portraits, the most core step is to apply appropriate "tags" to the users. These labels to the portraits are generated by analyzing the collected user data, and can be used for these The data information is highly summarized.

User portraits are also called user roles. As an effective tool for sketching target users, connecting user demands and design directions, user portraits have been widely used in various fields. We often use the most obvious and close word to life in the actual operation process to connects user attributes, behaviors and expectations. As a virtual representative of actual users, the user roles formed by user portraits are not constructed outside the product and the market. The user roles formed need to be representative and capable of Represent the main audience and target group of the product.

Li [8] and others combined the 38 million e-commerce activity data generated by Groupon with the combination of domain knowledge from e-commerce with data mining and graph theory methods to improve the user viscosity of the product; Lu [9] proposed an algorithm for topic interest mining of Sina Weibo users based on tags and two-way interactions. The algorithm for mining user topic interests through free tags and social interaction on Sina Weibo. Wikipedia expanded and expressed as vectors, and uses the tf-idf method to obtain the vector of each microblog. Then analyzes the sort of user interest tags according to the forward and the backward arrangement to get the final user interest tags; Punit [10] proposed a hybrid, set-based clustering algorithm that can use fast data space reduction and smart sampling strategies; Liu [11] proposed an improved clustering-based collaborative filtering recommendation algorithm introduces time decay function for preprocessing the user's rating and uses project attribute vectors to characterize projects, user interest vectors to users and use clustering algorithms to cluster the users and the projects respectively. The algorithm can portrait for users in multi dimension and reflect the user's interest changing. Jia [12] employed the telecom data and proposes a user clustering and influence power ranking scheme. The scheme is implemented through three stages, i.e. the user portrait analysis stage, the user clustering analysis stage and the ranking stage of user influence power; Cuiling [13] proposed a new multidimensional foreign language learning community, which dynamically evaluates and classifies students' learning status in real time to form student portraits, and automatically guides students to perform group learning at different levels of ability in the learning community. Choosing a reasonable and effective clustering algorithm successfully implemented the system's data mining application in the e-learning environment.

In the study of user portraits, we learned that generating new labels by clustering algorithms is a very common method in corporate user portraits, and the current user portrait technology is not suitable for college data. There are differences on purpose of college user portrait and enterprise user portrait. Student portrait is the production of college big data, which is applied to help administrator of college to know students, master student status by real-time and exactly locate abnormal crowed. Otherwise, enterprise portrait is more applied to deal with propensity to consume and propensity of topic.

From the current research on student portraits in universities, we can see that there are two major problems with the existing student portrait systems. On the one hand, many studies are only based on the system architecture of student portraits. Called for discussion without mentioning specific student portrait establishment methods; on the other hand, the existing student portrait system analysis has fewer dimensions, the breadth of the data source for analysis is not sufficient, and other dimensions of the student's information, such as behavior information, etc. Synthesize student portraits. In addition, the method of establishing student portrait tags is more based on the experience of higher education researchers. It does not have comprehensive representation and the data mining model is too simple to reflect the recent thinking of students in real time. Dynamic, and model accuracy is not high.

2.2 Neural Network Behavior Prediction

Donkers [14] extended the regression neural network by considering the unique features of the recommendation system domain. One of these features is a clear concept of user recommendations specifically generated for it, showing how to use a sequence of consumer items in a new type of gating loop unit. To represent individual users to effectively generate personalized recommendations for the next project; Lefebvre-Brossard [15] proposed a new method that relies on recurrent neural networks and word embeddings to match the problem of learners looking for guidance, and mentors willing to provide such guidance; Wang [16] proposed an end-to-end encrypted traffic classification method with one-dimensional convolution neural networks. This method integrates feature extraction, feature selection and classifier into a unified end-to-end framework, intending to automatically learning nonlinear relationship between raw input and expected output; Lin [17] designed a convolutional neural network with cross autoencoders to generate user-scope content attributes from low-level content attributes. Finally, they propose a deep neural network model to incorporate the two types of userscope attributes to detect users' psychological stress. Tommy [18] used one-dimensional convolution neural network and short-term memory neural network to mine user information on Facebook, so as to predict user personality. Cai [19] proposed a CNN-LSTM attention model to predict user intents, and an unsupervised clustering method is applied to mine user intent taxonomy.

From the above, we can see that when using deep learning to predict user behavior, convolutional neural networks and recurrent neural networks are often used. The advantage of convolutional neural networks is that they can store data in high dimensions and can introduce the time axis is used as the data dimension to solve the complexity of the data structure of college students. The advantage of the recurrent neural network is that the data itself has a requirement for the input sequence. It has a strong time correlation and can better connect students with Time is used as the dimension data during the school, such as teaching week, but in actual operation, we found that the current student data of colleges and universities cannot evaluate students in a short unit time.

One week's performance is evaluated, so it is impossible to implement classification prediction of student portrait tags with RNN, so we chose to use CNN as a component of the model for classification and prediction of portrait tags when building a hybrid neural network model.

3 Label Design of Student Portrait Based on High Dimensional Clustering

3.1 Clique Algorithm

Students in colleges and universities will generate all kinds of data, such as student basic data, teaching data, card consumption data, access control data, library borrowing data, gateway traffic, web browsing data, etc. These data can be divided into structures Data and unstructured data. Structured data is structured data. Generally, the amount of data is small, but it has good data characteristics. Unstructured data includes semi-structured data and unstructured data. The amount of data is huge and the characteristics of the data are not obvious. Using high-dimensional clustering technology, we can perform data mining, transform complex multi-source heterogeneous data into simple semantic labels, and lay the foundation for the establishment of student portraits.

The grid clustering algorithm has good scalability for data set size, can handle large-scale data sets, and the clustering results are not affected by the order of data input. It is suitable for the diverse, high-dimensional, and individual student data sources of college students in this paper. The characteristics of the grid clustering algorithm are more intuitive and easy to understand, and it can be used to classify student data in the management of college students in a practical way, and positively promote higher education research.

This paper plans to use clique-based clustering algorithm, that is, automatic subspace clustering algorithm, to cluster students' various data, and to explore the potential connections between students in static data and dynamic data, so as to establish student portrait labels. The model is more suitable to explore the degree of influence between various attributes of students, so as to determine the impact of specific attributes or behaviors on students.

Clique has the advantages of grid-like algorithms and is not sensitive to the order of data input. It does not need to assume any standardized data distribution. It scales linearly with the size of the input data. It has good scalability when the data dimension increases. For large databases The clustering of high-dimensional data in is very effective. The advantages of the above clique are very suitable for dealing with student data, the input of student historical data does not require order, and the data structure is highly designable, which is easy to expand and adjust.

The clique algorithm pseudo-code is shown in Algorithm 1.

Algorithm 1. The clique algorithm

Input: D_{k-1} (Set of all $k-1$ dimensional dense units);
Ouput: all k-dimensional candidate denses;
Algorithm:

      insert into S_k ;

      select $u_l[l_1,h_1], u_l[l_2,h_2], \ldots, u_l[l_{k-1},h_{k-1}], u_l[l_{k-2},h_{k-2}]$;

      from $D_{k-1}u_1, D_{k-2}u_2$;

      where $u_1a_1 = u_2a_1, u_1l_1 = u_2l_1, u_1h_1 = u_2h_1, u_1a_1 = u_2a_2, u_1l_2 = u_2l_2,$

      $u_1h_2 = u_2h_2, u_1a_{k-2} = u_2a_{k-2}, u_1l_2 = u_2l_2, u_1h_2 = u_2h_2, \ldots,$

      $u_1a_{k-2} = u_2a_{k-2}, u_1l_{k-2} = u_2l_{k-2}, u_1h_{k-2} = u_2h_{k-2}, u_1a_{k-1} < u_2a_{k-1}$;

The clique algorithm uses a fixed mesh division method to divide the data space equally according to different parameters entered by the user, and this will cause some data sample points in the sparse area at the class boundary that should belong to this clustering cluster to be considered non-compliant Dense units may be partitioned into different adjacent sections, which seriously damages the integrity of the original dense area.

According to the strategy of pruning the minimum description length, the entire data space is divided into multiple different subspaces. In each subspace, the density units are forcibly divided into n groups and the corresponding data is covered. Under this rule, the largest one can be found. Subspace, the remaining subspaces will be ignored. Under this strategy, some of the subspaces have been pruned, and the dense units you are looking for may be located in it, which leads to the incompleteness of the dense units. Although the method improves the operation efficiency of the algorithm, the lost subspaces affect the accuracy of the clustering results.

3.2   Auto-CLIQUE Algorithm

In high-dimensional data space, the number of grid cells will increase exponentially, and fixed grid division based on input parameters will likely cause the same cluster to be divided into multiple regions. The proposed discrete coefficients adopt adaptive meshing based on discrete coefficients and set iterative grid density thresholds to dynamically and flexibly divide the mesh to improve the processing effect and performance of the algorithm without the need for user input. Reliance on the knowledge of experiments to improve the quality of clustering.

Adaptive Meshing Based on Discrete Coefficients. In order to eliminate the influence of the level of data value and different measurement units on the measurement of the degree of dispersion, a dispersion coefficient is set, which is the ratio of the standard deviation of a set of data to its mean value. It is used to measure the relative degree of dispersion of the data, and does not require user input Parameters, reducing the algorithm's dependence on prior knowledge and improving the quality of clustering. The formula for the discrete coefficient is:

$$D_j = \frac{S_j}{\overline{X}_j} \tag{1}$$

where D_j is the dispersion of the j-dimensional data of the data set, S_j is the standard deviation of the j-th dimension, $\overline{X}_j$ is the mean of the j-dimensional data. The standardized dispersion formula is:

$$D_s = \frac{1}{1 + \frac{1}{d} * \sum_d^{j=1} D_j} \tag{2}$$

where D is the dimension of the data object, D_s the smaller the value of, the greater the dispersion of the data set, D_s the larger the value of, the smaller the discreteness of the data set. The definition of the segmentation parameter m is:

$$m = D_s * \sqrt[d]{N} \tag{3}$$

where N is the number of samples in the data set.

Grid Density Threshold Setting Based on Iteration. In addition, we also propose to use a recursive algorithm to calculate the data density threshold according to the data set itself, which can effectively separate the dense grid and the sparse grid, and once again reduce the dependence on prior knowledge. The density threshold formula is:

$$\rho[i+1] = \rho[i] + \frac{grid[i]}{data[i]} \tag{4}$$

where $\rho[i]$ is the ith density threshold, $grid[i]$ is greater than or equal to the density threshold $\rho[i]$ the number of grids, $data[i]$ for $grid[i]$ grids exceed density threshold $\rho[i]$ the total number of extra data points.

4 Classification Model of Student Portraits Based on Hybrid Neural Network

The recent labeling of student portraits is more time-efficient than the feature labeling a long time ago, but at the same time the meaning of labels generated by static data cannot be denied, so this paper plans to propose a method based on recurrent neural networks and feed forward neural networks. Hybrid neural network models to solve this situation.

4.1 Hybrid Neural Network Framework

In the process of establishing the student portrait label, we found that taking the student's "learning effort" label as an example, the label is not only affected by the performance results of the teaching affairs data in the static data, but also by the student's routine of the semester. The frequency and other behavioral effects of entering and leaving a school, that is, the labeling results are affected by both the long-term static properties of the students and the short-term behaviors. The two results must have different choices for the labeling results.

In this paper, the following improved one-dimensional convolutional neural network and BP pre-feedback neural network are combined with SotfMax and trained with real data. The network structure is shown in Fig. 1.

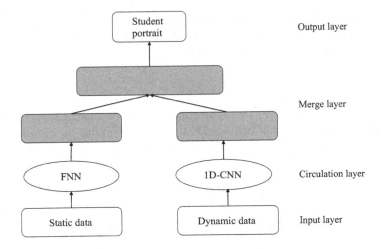

Fig. 1. Basic framework of hybrid neural network

4.2   Data Prediction Based on Feed Forward Neural Network

Compared with the student's dynamic behavior data, the static data composed of the student's basic information and teaching information is much easier to handle, but it represents the student's long-term behavioral trend, so we can use this kind of data to predict the student's long-term development trend. This paper uses a front-feedback neural network (FNN) to perform data mining on this part of the data. The BP neural network is used to determine the corresponding labels of the student portraits. An example is shown in Fig. 2.

We use a vector data structure to store training data and test data in the BP neural network. In addition to the basic information data of the students, it also includes relevant data about the last semester about campus life in the campus: average dining hall consumption, Standard deviation, number of consumptions; average consumption, standard deviation, and number of times of consumption in supermarkets; average consumption, standard deviation, and number of times of consumption in online schools; average number of consumption, averages of bathroom recharges in schools, and educational data since students enrolled: all courses Average scores, weighted scores, and standard deviations of scores, as well as the average scores of public basic compulsory courses, practical compulsory courses, general education and school elective courses, subject basic compulsory courses, subject basic elective courses, professional optional courses, and professional limited elective courses, Weighted score, standard deviation of scores. The data of each student is shown in Table 1.

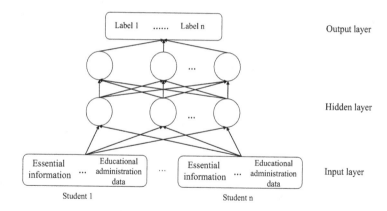

Fig. 2. Basic framework of hybrid neural network

Table 1. FNN student data sheet

Type of data	Data breakdown	Data field
Living data	Canteen consumption	Mean, standard deviation, Consumption
	Supermarket Consumption	Mean, standard deviation, Consumption
	Recharge the school bathroom	Consumption times, average
Academic data	All courses	
	Public Foundation Compulsory Course	Average score,
	Practical Compulsory Course	
	General Education and School Electives	Weighted score,
	Required Subject Courses	
	Subject Elective Courses	Fractional standard deviation
	Professional optional Courses	
	Restricted Courses	
Basic information data		

To ensure accuracy, we set up four-layer neurons for the BP neural network in this paper. The activation function uses the relu activation function mentioned above. The network structure is shown in Table 2.

Table 2. FNN data structure table

Layer (type)	Output Shape	Param #
Input	(None, 26)	0
Dense1	(None, 108)	2916
Dense2	(None, 52)	5668
Dense3	(None, 12)	636
Dense4	(None, 6)	78

Dynamic Data Prediction Based on One-Dimensional Convolutional Neural Network (1D-CNN). Compared to the static data of students, the dynamic data of students has a lot of time-oriented behavioral data, which is more complicated. In it, we find that the dynamic data of students is different from the static data in two data formats, which are integrated into The unified data format will cause a large amount of compression and loss of data, or there will be a large amount of data redundancy. For student behavior data, we clean and organize the data and store it in a matrix. The data is as follows as shown in Table 3.

Table 3. 1D-CNN student data sheet

Student serial number	Card data item	Time (teaching week)			
Student n	Canteen Consumption	First 1 week	First 2 week	...	First 20 week
	Canteen consumption standard deviation				
	Total supermarket consumption				
	Supermarket consumption standard deviation				
	Total online consumption				
	Total bathroom consumption				

Dynamic data exists in the form of a two-dimensional matrix. In deep learning, convolutional neural networks have always studied in depth two-dimensional data such as video, images, and audio. One-dimensional convolutional neural networks are suitable for sequence data or languages. Data, so we borrowed its way of processing two-dimensional data and used one-dimensional convolutional neural networks to process student dynamic data.

The structure of a common one-dimensional convolutional neural network model is: an input layer, a convolutional layer, a pooling layer, a three-layer fully connected layer, and an output layer. The convolutional layer can be expressed as:

$$C = f(xk + b) \qquad (5)$$

where x represents the input, k represents the convolution kernel, and b represents the offset value. f is the activation function. Common activation functions include relu, tanh, and sigmoid. The relu function is used in this paper (see Eq. (6)).

$$f(x) = max(0, x) \qquad (6)$$

The convolution layer C performs sliding on the serialized data set and convolves with the original data to obtain the feature layer.

Pooling layer S, the pooling layer refers to the down sampling layer, which combines the output of a cluster of neurons in the previous layer with a single neuron in the lower layer. The pooling operation is performed after non-linear activation, where the pooling layer helps reducing the number of parameters and avoiding over fitting, it can also be used as a smoothing method to eliminate unwanted noise. The S layer can be expressed as:

$$S = \beta \cdot (C) + b \tag{7}$$

where β and b are scalar parameters, and $down$ is a function selected by down sampling. There are an average pooling layer and a maximum pooling layer. This paper uses both methods to improve the functionality of the perception area.

The output layer uses a $Softmax$ function classifier, assuming the output of a one-dimensional convolutional neural network is $y_1, y_2, ..., y_n$, and the output after $Softmax$ layer is:

$$Softmax\left(y\right)_i = \frac{e^{y_i}}{\sum_n^{i=1} e^{y_j}} \tag{8}$$

The one-dimensional convolutional neural network model designed according to the data mining needs of college students in this paper is shown in Fig. 3.

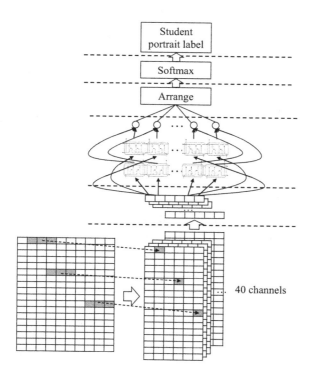

Fig. 3. Improved 1D-CNN model structure

The data structure of each layer in 1D-CNN is shown in Table 4.

Table 4. 1D-CNN data structure table

Layer (type)	Output Shape	Param #
Reshape	(None, 18, 8)	0
Conv1D	(None, 16, 40)	1000
MaxPooling1D	(None, 8, 40)	0
LSTM1	(None, 8, 80)	38720
LSTM2	(None, 8, 20)	8080
AveragePooling1d	(None, 20)	0
Dense	(None, N)	42

Between the largest pooling layer and the average pooling layer, we have introduced two layers of long-term and short-term memory networks (LSTM). Using the recurrent neural network's memory and parameter sharing features, it improves the connection between the student portrait label and the upper and lower teaching week. The relevance of making predictions.

The LSTM network is a time-recurrent neural network, which is specially designed to solve the long-term dependency problem of general RNN (recurrent neural network).

In the text $x_1, x_1, \ldots, x_n$ represents the output of the previous MaxPooling layer, with a structure of 40 channels in 1 row and 8 columns. x_n corresponds to the n-th column (n maximum 8) array, this array includes 40 eigenvalues, $h_1, h_1, \ldots, h_n$ express $x_1, x_1, \ldots, x_n$ after each output of the cell, when the LSTM structure is 80 channels in 1 row and 8 columns, h_n express x_n (40 eigenvalues) after the cell output (80 eigenvalues). A schematic reference is shown in Fig. 4.

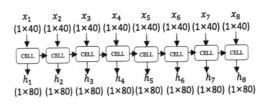

Fig. 4. LSTM layer data structure in 1D-CNN

The training objective function uses a cross-entropy formula function with the following formula:

$$L = -[y log\hat{y} + (1 - y) \log (1 - \hat{y})] \tag{9}$$

In Eq. (9), when $y = 1$, $L = -\log \hat{y}$ the closer the predicted output is to the true sample label 1, the smaller the loss function L is; The closer the predicted output is to 0, the larger L is. Therefore, the change trend of the function completely meets the actual needs; when $y = 0, L = -\log(1 - \hat{y})$ the closer the predicted output is to the true sample label 0, the smaller the loss function l; The closer the predicted function is to 1, the larger l, the change trend of the function also fully meets the actual needs.

5 Experiment

5.1 Data Set and Experimental Environment

Window version: Windows 10 64-bit operating system Hardware environment: i7-6700HQ CPU 2.60 GHz; RAM 8.00 GB Anaconda3 running environment: Python 3.6.0; Python Package numpy 1.15.2, pandas 0.25.1, tensorflow 1.14.0, keras 2.24, scikit-learn 0.21.3, matplotlib 3.1.1, seaborn 0.9.0.

5.2 Model Evaluation Criteria

Confusion Matrix. As shown in Table 5, for the prediction of obtaining scholarship, the examples can be divided into true examples (True Positive (TP)), false positives (FP), true Negative (TN), and False Negative (FN) according to the combination of their real categories and algorithm recognition. We combine the above to form a "confuse matrix of predicting scholarship".

Table 5. Confusion matrix of predicting scholarship

Reality	Forecast result	
	Obtain scholarship	Not acquired scholarship
Obtain scholarship	TP (real example)	FN (false counterexample)
Not acquired scholarship	FP (false positive)	TN (true and negative)

The true example is the correct choice; the false positive example is the wrong choice, which indicates a misjudgment, and the accuracy of the result is related; the false negative example indicates the missing data, which is related to the recall rate; the true and negative examples are not needed in this paper and will not be discussed.

Equation (10) represents the accuracy rate. The ratio of the number of TP in the recognition results found in a single run of the algorithm is as follows:

$$Precision = \frac{TP}{TP + FP} = \frac{TP}{all\,detections} \tag{10}$$

Accuracy, Recall and F1-Score Indicators. Equation (11) represents the recall rate. The ratio of the number of TP found in a single run of the algorithm to the number of all positive samples in the sample. The expression is:

$$Recall = \frac{TP}{TP + FN} = \frac{TP}{all\ ground\ truths} \tag{11}$$

Equation (12) represents the F1-Score indicator, which combines the results of $Precision$ and $Recall$. The value of $F1 - Score$ ranges from 0 to 1, where 1 represents the best model output, and 0 represents the model output result, worst. The expression is:

$$F1 - Score = 2 * \frac{Precision * Recall}{Precision + Recall} \tag{12}$$

5.3   Experimental Results and Analysis

With regard to the hybrid neural network model mentioned above, we selected 1,055 undergraduates in a certain college and university to train the model, which can be used to determine whether there is an academic crisis for students, whether they are potentially poor students, whether there are psychological warnings, It is expected that scholarships will be obtained and recent graduates will become outstanding graduates.

BP Neural Network Model Test. Taking the expected scholarship for students as an example, the results of the cross-entropy loss function during the training of the BP neural network model using the corresponding data are shown in Fig. 5.

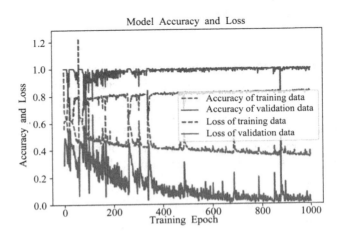

Fig. 5. Cross entropy loss function in the training process of BP neural network model

The related test parameters of the BP neural network model are shown in Table 6.

Table 6. Comparison of four neural networks validation parameters

Type	Data type		Precision		Recall		F1-score		Time	Accuracy
	Static	Dynamic	0	1	0	1	0	1	(us/step)	
Common 1D-CNN		✓	0.70	0.41	0.85	0.66	0.75	0.24	42	0.66
Improved 1D-CNN		✓	0.73	0.40	0.72	0.41	0.72	0.41	126	0.62
BP neural network	✓		0.86	0.83	0.94	0.67	0.90	0.74	27	0.85
Hybrid neural network	✓	✓	0.88	0.78	0.91	0.74	0.89	0.76	160	0.85

ps: 0 in the table means no scholarship (722 supports), 1 means scholarship (333 supports)

From the chart above, we can see that the *accuracy* of the BP neural network model in this paper is 0.85, the $F1 - score$ of the un-scholarship is 0.90, and the F1-score of the scholarship is 0.74. The results of scholarships are obtained for prediction. The main reason for the better performance of this model is that the training data contains the student's various achievements, which can affect the possibility of students receiving scholarships to a large extent.

Improved 1D-CNN Neural Network Model Test. In the improved 1d-cnn neural network model test, we used a common 1d-cnn neural network model with a similar structure to compare with it. In addition, we implement it according to the one-dimensional neural network model proposed by Waibel A [20] , and compare it with other neural network models in this paper. The network structure is shown in Table 7.

Table 7. Common 1D-cnn data structure table

Layer (type)	Output Shape	Param #
Reshape	(None, 18, 8)	0
Conv1D	(None, 16, 40)	1000
MaxPooling1D	(None, 8, 40)	0
AveragePooling1D	(None, 20)	0
Dense	(None, N)	42

This section of the experiment uses the school life data of the students mentioned in the 2018–2019 school year as input, and each student's data is stored in a matrix in the form of Table 3. The parameter settings are as follows: the kernel size is set to 3, Use the *EarlyStopping* function as a callback function, set vac_{loss} to monitor, and set the value of the patience parameter to 200, which means that during the training process, when the loss value of the test set is constant for 200 trainings, stop training in advance. Training batch size (batch size) is set to 256, the learning rate is set to 0.01, and the maximum number of iterations (epoch) in the training phase is set to 2000.

The two 1D-CNN model pairs are shown in Table 6. From Table 6, we can see that the improved 1D-CNN model has improved the classification accuracy rate, classification recall rate, $F1 - Score$, and $accuracy$ rate of whether students will receive scholarships, especially for "getting scholarships" Relevant parameters of classification, because in the data set, the number of students corresponding to the "not awarded scholarship" label is higher than "received scholarship", so the improvement of this type of parameters indicates that the improved 1D-CNN model is more common than the ordinary 1D-CNN model The accuracy has been greatly improved. However, as the improved 1D-CNN model adds two LSTM layers to enhance the connection of student life data during the teaching week, the time complexity has increased. It takes time from 42 us to 126 us.

The results of the cross-entropy loss function training using the improved 1d-cnn model in this paper are shown in Fig. 6.

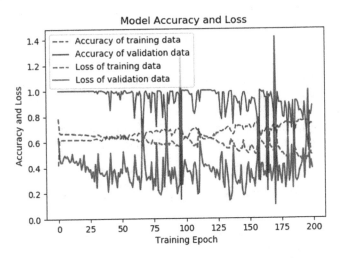

Fig. 6. Cross entropy loss function in the training process of improved 1D-CNN model

Hybrid Neural Network Model Test. The cross entropy loss function during the training of the hybrid neural network model mentioned in this paper is shown in the Fig. 7. The overall trend is that the accuracy of the training data and test data gradually approaches 1, and the loss gradually decreases. The training convergence process realizes the connection of static and dynamic data of students, and can mine the potential relationship between static data such as student performance in various subjects and dynamic data generated from campus life, and can better predict the situation of students receiving scholarships. Provide guidance for college student management.

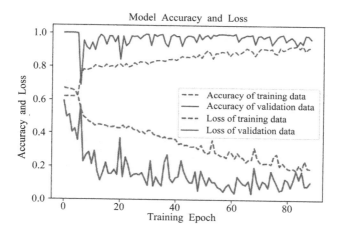

Fig. 7. Cross entropy loss function in the training process of hybrid neural network model

The relevant test values are shown in Table 6. From the table above, we can see that in the hybrid neural network model, students have higher accuracy in the prediction of not receiving the scholarship label, and have certain prediction capabilities in obtaining the scholarship label, and the overall accuracy can be correct. Predicting whether students will receive scholarships has achieved our goal of building a hybrid neural network.

6   Conclusion

This paper builds a college student portrait system based on an improved one-dimensional convolutional neural network and a front-feedback neural network. Among them, an optimization is proposed in the establishment of a one-dimensional convolutional neural network model. The 1d-cnn maximum pooling layer, the LSTM layer and the average pooling layer strengthen the connection of college student data in the unit of teaching week.

In addition, more information is provided on student personal data mining, predicting students' future behavioral tendencies, so as to implement precision poverty alleviation, academic early warning, psychological early warning and other tasks in efficient work.

The future development direction is to implement the application of algorithms based on the specific status of universities, establish a large database of student behavior information, establish a student information IoT acquisition system, and establish a student information distributed system, so as to improve data collection, data cleaning, and real-time prediction functions.

References

1. Zhou, Q., Mou, C., Yang, D.: Research progress on educational data mining: a survey. J. Softw. **26**(11), 3026–3042 (2015). (in Chinese)
2. Castillo, G., Gama, J., Breda, A.M.: Adaptive Bayes for a student modeling prediction task based on learning styles. In: Brusilovsky, P., Corbett, A., de Rosis, F. (eds.) UM 2003. LNCS (LNAI), vol. 2702, pp. 328–332. Springer, Heidelberg (2003). https://doi.org/10.1007/3-540-44963-9_44
3. Pandey, M., Sharma, V.K.: A decision tree algorithm pertaining to the student performance analysis and prediction. Int. J. Comput. Appl. **61**(13), 1–5 (2013)
4. Yuan, H., Xu, W., Wang, M.: Can online user behavior improve the performance of sales prediction in E-commerce?. In: IEEE International Conference on Systems. IEEE (2014)
5. Alahi, A., Goel, K., Ramanathan, V., et al.: Social LSTM: human trajectory prediction in crowded spaces. In: 2016 IEEE Conference on Computer Vision and Pattern Recognition (CVPR). IEEE (2016)
6. Wong, B.K., Selvi, Y.: Neural network applications in finance: a review and analysis of literature (1990–1996). Inf. Manage. **34**(3), 129–139 (1998)
7. Brickey, J., Walczak, S., Burgess, T.: Comparing semi-automated clustering methods for persona development. IEEE Trans. Softw. Eng. **38**(3), 537–546 (2012)
8. Li, K., Deolalikar, V., Pradhan, N.: Mining lifestyle personas at scale in e-commerce. In: IEEE International Conference on Big Data. IEEE (2015)
9. Deng, L., Jia, Y., Zhou, B., et al.: User interest mining via tags and bidirectional interactions on Sina Weibo. World Wide Web-internet Web Inf. Syst. **21**(2), 515–536 (2018)
10. Punit, R., Dheeraj, K., Bezdek, J.C., et al.: A rapid hybrid clustering algorithm for large volumes of high dimensional data. In: IEEE Transactions on Knowledge and Data Engineering, p. 1 (2018)
11. Xiaojun, L.: An improved clustering-based collaborative filtering recommendation algorithm. Cluster Comput. **20**(2), 1281–1288 (2017)
12. Jia, Y., Chao, K., Cheng, X., et al.: Big data based user clustering and influence power ranking. In: 2016 16th International Symposium on Communications and Information Technologies (ISCIT). IEEE (2016)
13. Lv, C.: Application study on data mining technology of English learning virtual community. In: International Conference on Intelligent Transportation. IEEE Computer Society (2018)
14. Donkers, T., Loepp, B., Jürgen, Z.: Sequential user-based recurrent neural network recommendations. In: Eleventh ACM Conference on Recommender Systems. ACM (2017)
15. Lefebvre-Brossard, A., Spaeth, A., Desmarais M.C.: Encoding user as more than the sum of their parts: recurrent neural networks and word embedding for people-to-people recommendation. In: Conference on User Modeling. ACM (2017)
16. Wang, W., Zhu, M., Wang, J., et al.: End-to-end encrypted traffic classification with one-dimensional convolution neural networks. In: 2017 IEEE International Conference on Intelligence and Security Informatics (ISI). IEEE (2017)
17. Lin, H., Jia, J., Guo, Q., et al.: User-level psychological stress detection from social media using deep neural network (2014)
18. Tandera, T., Hendro, Suhartono, D., et al.: Personality prediction system from Facebook users. Procedia Comput. Sci. **116**, 604–611 (2017)

19. Cai, R., Zhu, B., Ji, L., et al.: An CNN-LSTM attention approach to understanding user query intent from online health communities. In: 2017 IEEE International Conference on Data Mining Workshops (ICDMW). IEEE (2017)
20. Waibel, A., Hanazawa, T., Hinton, G., et al.: Phoneme recognition using time-delay neural networks. IEEE Trans. Acoust. Speech Signal Process. **37**(3), 328–339 (2002)

Efficiently Detecting Light Events in Astronomical Temporal Data

Chenglong Fang[✉], Xieyang Wang, Jianqiu Xu, and Feng Wang

Nanjing University of Aeronautics and Astronautics, Nanjing, China
{fangcl,xieyang,jianqiu,marvel_agent}@nuaa.edu.cn

Abstract. Detecting light events is a crucial problem in astronomical temporal data, the task of which is to find gravitational microlensing and flare star among the Milky Way. We propose an efficient method called FLMM (Fast Locating based on Median and Mean) to detect light events. The idea is to determine the location of suspicious light events in astronomical data by means of median-mean value and peak value in data after preprocessing. We combine the respective advantages of FLMM and Dynamic Time Warping (DTW) to improve the recognition accuracy. In order to balance between efficiency and accuracy, we use DTW to match the raw data of the suspicious light event. The experimental results demonstrate that for the processing of 0.93 million pieces of historical data, the combined method is nearly 3 times faster than DTW. The processing accuracy of abnormal data is improved by 2 orders of magnitude.

Keywords: Light event · Anomaly detection · Median-mean · Temporal data

1 Introduction

In recent years, the continuous progress of science and technology has aroused great concern for the hidden potential value of big data. Temporal data is constantly produced all the time. The support of large-capacity storage devices makes the amount of data in various research fields increase exponentially, which brings great challenges to the analysis of data with great value in large-scale data [5].

In this paper, we study effectively detecting light events of rare celestial bodies such as gravitational microlensing [9] and flare star in astronomical temporal big data. Microlensing is of great significance to the development of astronomy and the study of the Milky Way [8]. At present, microlensing has been widely used in the study of the Milky Way to provide a basis for the study of some planets that are hard to be observed. Figure 1 shows the light curve characteristics of gravitational microlensing and flare star in the real data set. The X-axis represents the astronomical Julian day and the Y-axis represents the magnitude of the celestial body.

© Springer Nature Switzerland AG 2021
X. Meng et al. (Eds.): SpatialDI 2020, LNCS 12567, pp. 184–197, 2021.
https://doi.org/10.1007/978-3-030-69873-7_13

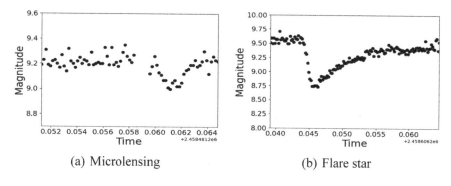

(a) Microlensing

(b) Flare star

Fig. 1. Light curve

With the development of modern astronomical observation technology, the amount of data is increasing rapidly at the magnitude of TB and even PB. With the support of big data, time-domain astronomical observation technology has made great progress and a number of exoplanets have been discovered. Some theories predict that there are plenty of "special planets" in the Milky Way, but finding them remains a bottleneck [10]. The ability to discover such rare and small-probability astronomical events accurately and quickly in large-scale data has became an urgent problem. The smaller the mass of some of the "special planets" being explored, the shorter the light event curve time of the microlensing event. As a result, we need high temporal resolution and large sky coverage from the observation. By collecting a batch of time-domain astronomical observation data with high time resolution, the discovery efficiency of this rare event is improved. We use data from the Ground-based Wide Angle Camera (GWAC) [13] at the National Astronomical Observatories of the Chinese Academy of Sciences. GWAC is a time-domain astronomical survey equipment constructed specially, which can obtain a sampling observation data every 15 s. Due to the wide field coverage ability of GWAC, millions of light curve samples with 15-second sampling resolution have been obtained. The data provides a possible basis for the discovery of "special planets" with Earth-like masses on the hourly scale. The research presents two major challenges to the discovery of rare celestial light events in large astronomical data: (i) efficiency, and (ii) accuracy.

Efficiency. Efficiently detecting light events rapidly in large scale astronomical temporal data is a challenging issue [5]. Iteratively processing the data one by one is very time-consuming and expensive. Therefore, we propose FLMM (Fast Locating based on Median and Mean) to detect the suspicious cases of light events. In this way, some data that does not need to be detected can be excluded directly for comprehensive analysis of some possible points.

Accuracy. In order to effectively detect the light events of rare celestial bodies from astronomical big data, we first need to consider the error of data collection. Due to the error of equipment and the influence of weather, data inconsistency will inevitably occur. Since errors exist in data points, the noise brought by large-scale data will have a huge impact on the analysis of spatial and temporal data. The geographical environment and climate factor are very important to the noise removal in data sampling. The roughness of the original data not only affects the efficiency, but also the accuracy. The smooth denoising of Moving Average (MA) [1] model will make the data with large errors in the original data rise smoothly, and then be mistaken as rare light event to affect the accuracy of detection results. In the aspect of data denoising, we utilize the method Wavelet Transform (WT). Compared with Short-Time Fourier Transform (STFT), WT not only inherits and develops the idea of STFT localization, but also overcomes STFT window size without changing with frequency [16].

Removing the noise of amounts of data is a time-consuming task. First of all, we find out the suspicious points through the rapid positioning of the data then carry out denoising for the data with suspicious light events. Finally we exclude the data without suspicious points. In order to effectively identify the light event of celestial bodies, we combine FLMM with Dynamic Time Warping (DTW) similarity matching algorithm to improve the accuracy of detection results. We further exclude the abnormal data of the preliminary results by matching the corresponding raw data and solve the misjudgment result of rare celestial light event caused by certain data preprocessing error.

We use the real data collected by GWAC and the dataset contains approximately 0.93 million astronomical historical data. We perform the evaluation among FLMM, DTW and FLMM-DTW. The experimental results demonstrate that (i) for the processing of 0.93 million historical data, FLMM-DTW runs nearly three times faster than DTW matching; (ii) the processing accuracy of abnormal data is improved by two orders of magnitude. Considering the special value of the original data and the noise data, we substitute the original data of the results into the algorithm for relatively rough matching, thus we further improving the accuracy of the data analysis.

The rest of the paper is organized as follows. The related work is discussed in Sect. 2. We make the definitions of light events of celestial in Sect. 3. FLMM and FLMM-DTW are introduced in Sect. 4. Section 5 presents our experimental results. We conclude the paper in Sect. 6.

2   Related Work

In the literature, there is a lot of related work on temporal data processing including the query and maintenance of temporal data in multiple fields. Some papers [17,18] examine the problem of computing temporal aggregates over data streams and propose to maintain temporal aggregates dynamically and progressively. The paper [17] provides a useful trade-off between aggregation detail and storage space. In terms of querying large-scale temporal data, these papers [7,15] analyze the efficiency of typed top-k query effectively and propose query algorithms to support point, interval and continuous queries. The paper presents a simple indexing scheme that supports temporal ranking queries efficiently [7]. At the same time, the index structure and query results are well visualized implemented in [14]. The paper examines not only the current snapshot of the data but also the historical and considers durable top-k queries [6], which looks for objects whose values are among the top k for at least some fraction of the times during a given interval. In terms of processing interval timestamp data, the paper [4] proposes relational algebra to provide native support for the three attributes of sequence semantics. For possible timestamp error, *Song et al.* proposes a method for cleaning timestamps [11].

Effective anomaly detection and repair for temporal data have been extensively investigated. To detect and clean time series data, speed constraint-based method is proposed by the paper [12]. Using smoothing-based models such as MA for data denoising and cleaning in temporal data will make data cleaning too smooth. Cleaning this model makes some data with excessively large abnormal values smoothed to produce more abnormal data. In the study on the characteristics of temporal data, if entities represented by the record change over time, approaches that use temporal information may do better than approaches that do not [3]. In [3], by focusing on the probability of a given attribute value reappearing over time, an entity may change the property value in a way depending on the past value of the entity.

We have done some research on some features of temporal data. We divide the data according to the time interval feature, which does not destroy the features inside the temporal data. To analyze user behavior over time, it is useful to group users into cohorts [2].

3   Preliminary

3.1   Problem Statement

Light events of rare celestial bodies detected in big data of astronomical time are mainly microlensing and flare star. Magnitude is a measure of celestial bodies luminosity. The brighter the brightness, the lower the magnitude, the darker the brightness, the greater the magnitude. Since both the microlensing and the flare star are accompanied by the increase and decrease in the brightness of celestial bodies over a short period of time, and we will explain microlensing and flare star together. There are some subtle differences in the brightness of celestial bodies, and we will detect microlensing and flare star separately.

Definition 1. *Suspicious points*
Let $M = <(m_1, t_1), ..., (m_n, t_n)>$, where $m_i \in \mathcal{R}$ and t_i is Instant, α is threshold such that $\exists [i, j] \subset [1, n], m_i > ... > m_j \vee \exists k \in [1, n], m_k < \alpha$.

If there are some points on M that are smaller than the threshold α or the magnitude fluctuation drops within a period of time. We treat these points as suspicious points.

Definition 2. *Gravitational microlensing candidate*
Let $M = <(m_1, t_1), ..., (m_n, t_n)>$, where $m_i \in \mathcal{R}$ and t_i is Instant, T_e is the time symmetry error, M_e is the magnitude symmetry error such that $\exists m_x \in M$: (i) $m_1 > ... > m_x \wedge m_x < ... < m_n$; (ii) $|(t_x - t_1) - (t_n - t_x)| \leq T_e$; (iii) $\forall k \in [1, x], |(m_{x-k} - m_x) - (m_{x+k} - m_x)| \leq M_e$.

With the increase of time, the corresponding magnitude M of the celestial body has an obvious phenomenon of decline and rise in a period of time and presents an axial symmetry relation in this phenomenon, so we can preliminarily judge the event as a candidate for gravitational microlensing.

Example 1. Let $M = <(9, 1), (9, 2), (9, 3), (7, 4), (5, 5), (4, 6), (5, 7), (7, 8), (9, 9), (9, 10)>$. The light curve features of microlensing are shown in Fig. 2(a). The red curve is our concern. To facilitate observation, we invert the Y-axis of the data to form a curve similar to the normal distribution, as shown in Fig. 2(b).

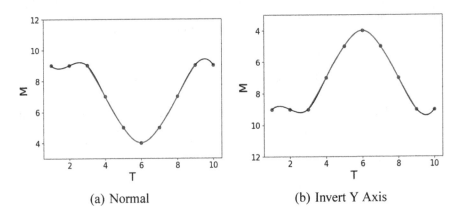

(a) Normal   (b) Invert Y Axis

Fig. 2. Characteristics of microlensing

Definition 3. *Flare star candidate*
Let $M = <(m_1, t_1), ..., (m_n, t_n)>$, where $m_i \in \mathcal{R}$ and t_i is *Instant*, T_e is the time symmetry error, M_e is the magnitude symmetry error such that $\exists\ m_x \in M$: *(i) the same as Definition 2 (i); (ii)* $(t_n - t_x) - (t_x - t_1) > T_e$.

If the sharp decline of the magnitude M in a certain time is followed by a slow rise and the time for the decline of M is significantly less than the rise time, then we can preliminarily judge the event as the flare star candidate.

Example 2. Let $M = <(9, 1), (9, 2), (5, 3), (4, 4), (5, 5), (6, 6), (7, 7), (8, 8), (9, 9), (9, 10)>$. The light curve features of flare star are shown in Fig. 3(a). We also pay attention to the red part of the curve and reverse the Y-axis to form a curve similar to the positive skewed distribution, as shown in Fig. 3(b).

If there are more than one similar microlensing or flare star over a period of time, this phenomenon is usually caused by a variable star, which is not included in our detection range and needs to be eliminated.

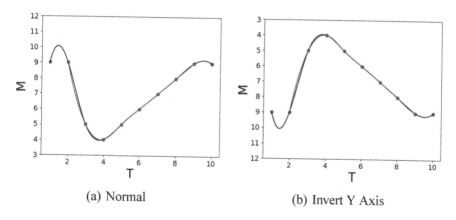

(a) Normal   (b) Invert Y Axis

Fig. 3. Characteristics of flare star

3.2 DTW Algorithm

DTW describes the corresponding relation between test data and reference template data by using a time warping function which satisfies certain conditions and obtains the similarity degree of the minimum accumulative distance between them. In order to determine whether there is data similar to the light event in astronomical data, we first design the data template according to the characteristics of microlensing and flare star. DTW is used to calculate the degree of similarity between time series, and the data satisfying the given threshold of similar distance is regarded as the candidate of light event.

Due to the continuity of astronomical temporal data and the huge scale, the error of template similarity matching for longer time series is very large. Since DTW is simply scaled linearly, short sequences are linearly enlarged to the same length as long sequences for comparison or long sequences are shortened to the same length as short sequences for comparison. Therefore, after the linear scaling of the data, the mismatch between the data will obviously reduce the similarity between the two sequences. In order to better match the template data and the pre-processed data, we split the data in terms of time to avoid the error effect of matching the shorter sequence through the longer time series.

4 Search by Suspicious Points

4.1 Suspicious Points Location

Due to the large scale of data, a large number of sub-sequences will be formed after the segmentation according to the time continuity. In order to improve operating efficiency, we extract the suspicious points in the data and exclude the rest points.

Median-Mean. Considering the time cost of data processing, we propose a fast locating method based on median and mean to determine whether there are suspicious celestial light events in sub-sequences. Let $M = <(m_1, t_1), ..., (m_n, t_n)>$ be the temporal data whose magnitude varies over time. Sort M by m_i from small to large by the sorting algorithm to get $M' = <(m'_1, t'_1), ..., (m'_n, t'_n)>$. The median and mean values in the time series are calculated by formulas (1) and (2). Set α as the threshold value of the minimum range of the change of magnitude of the light event. The data in M that differs from the median or mean by more than α are extracted as suspicious points and the suspicious points of microlensing and flare star are located respectively as shown in Fig. 4. For the data with multiple suspicious points, the data of the same group without suspicious points or the number of continuous segments greater than 2 after the segmentation is considered as an abnormal phenomenon and will be excluded.

$$m_{median} = \begin{cases} m'_{(n+1)/2} , n \text{ is odd} \\ (m'_{n/2} + m'_{n/2+1})/2 , n \text{ is even} \end{cases} \tag{1}$$

$$m_{mean} = \sum_{i=1}^{n} m_i/n \tag{2}$$

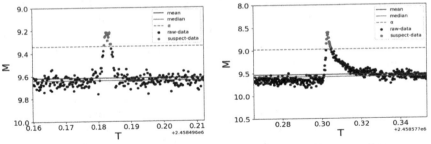

(a) Suspicious points of microlensing   (b) Suspicious points of flare star

Fig. 4. Locating suspicious data points by median and mean

Sliding Window. The characteristics of microlensing and flare star are the increase in brightness and the decrease in magnitude. Let W be the size of the sliding window, T_p be the period of the light event, $W \in [T_p / 2, T_p]$. Since the brightness of celestial objects fluctuates, the magnitude of the light event also fluctuates down. By detecting whether there is a magnitude fluctuating in the sliding window, the duration length of the drop satisfies the window width of $1 / \lambda$, and the data satisfying the conditions is marked as suspicious points. In a sliding window, as shown in Fig. 5, a continuously increasing wave is marked as suspicious points. The exception removal of suspicious points is the same as in Sect. 4.1. Sliding window makes up for some small light events missing from the median-mean method.

4.2   Light Events Detection

FLMM Algorithm. The data with the smallest magnitude in each segment of the suspicious points is taken as the vertex of the suspicious light event and substituted into the original data for regression calculation. The algorithm is given in Algorithm 1. We calculate the time covered by the points on both sides when the magnitude values on the left and right sides of the vertex return to the mean or median. Although the light event of the celestial body is relatively short, there is a certain time interval. Let t_l be the time covered by the point on the left, t_r be the time covered by the point on the right, T_p be the minimum period of light event, T_e be the symmetric time error on both sides of the microlensing. Suspicious points in each segment are classified into the following three cases: (i) Microlensing. As shown in Fig. 6(a), if the sum of the time satisfying the left and right coverage points is greater than the minimum period of the light event and the time difference is less than the symmetric time error, that is, $t_l + t_r \geq T_p \wedge |t_l - t_r| \leq T_e$ is regarded as the candidate of microlensing; (ii) Flare Star. In the case that the sum of time is greater than the minimum period and the difference between the right time and the left time is greater than the symmetric error, the phenomenon that satisfies $t_l + t_r \geq T_p \wedge t_r - t_l > T_e$ is regarded as the candidate of flare star in Fig. 6(b); (iii) Abnormal Phenomenon. If the sum of the time covered by both points is less than the minimum period or the difference between the time on the left and the time on the right is greater than the symmetric error, it meets $t_l + t_r < T_p \vee t_l - t_r > T_e$, and the condition can be removed.

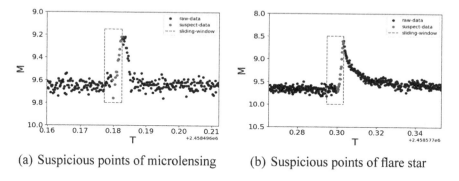

(a) Suspicious points of microlensing   (b) Suspicious points of flare star

Fig. 5. Locating suspicious data points by sliding window

FLMM-DTW Algorithm. Due to the fact that DTW is directly used for similar matching of data, a large number of useless matches will be generated, resulting in significantly lower operating efficiency and detection effect of the algorithm. We combine FLMM and DTW, which not only improves operating efficiency but also improves the accuracy of detection results. Let D be the template data where D_m is the template data of microlensing and D_f is the

---

Algorithm 1. *FLMM*

Input: M, α, T_p, T_e
Output: M', *type*
1: $m_{median} \leftarrow median(M)$
2: $m_{mean} \leftarrow mean(M)$
3: **for all** $(m_i, t_i) \in M$ **do**
4:     **if** $m_{median} - m_i > \alpha \vee m_{mean} - m_i > \alpha$ **then**
5:         $M' \leftarrow M' \cup (m_i, t_i)$
6: $M'' \leftarrow M'$ *is divided according to the time continuity of* t_i'
7: **if** $|M''|.length > 2$ **then**
8:     **return** *null*
9: **else**
10:     $t_l \leftarrow$ *calculate the time covered by the left point*
11:     $t_r \leftarrow$ *calculate the time covered by the right point*
12:     **if** $t_l + t_r \geq T_p \wedge |t_l - t_r| \leq T_e$ **then**
13:         **return** M'', *microlensing*
14:     **if** $t_l + t_r \geq T_p \wedge t_r - t_l > T_e$ **then**
15:         **return** M'', *flarestar*
16:     **return** *null*

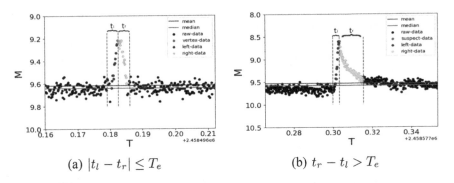

(a) $|t_l - t_r| \leq T_e$ (b) $t_r - t_l > T_e$

Fig. 6. Cover time comparison

template of flare star. We give the overall algorithm of FLMM-DTW in Algorithm 2. FLMM is used to locate the suspicious points and compare the left and right coverage time to determine the type of light events. DTW matching is used to reduce the result of error detection so as to improve the accuracy of detection. We improve the efficiency of DTW by reducing the number of template matching. Figure 7(a) and 7(b) show DTW similarity matching using template data in the curve fitting of microlensing candidate and the similarity matching of flare star candidate in Fig. 7(c) and 7(d). Let d_{max} be the maximum threshold of similar matching distance. The warping function is used to obtain the cumulative minimum distance of the data match and compare the cumulative minimum distance with the maximum distance threshold to further screen the candidates of light events. If the cumulative minimum distance is less than d_{max}, keep the candidate of light events, otherwise remove the candidate.

Algorithm 2. *FLMM-DTW*

Input: M, α, T_p, T_e, D, d_{max}
Output: *type*
1: **if** $FLMM(M, \alpha, T_p, T_e)$ is *null* **then**
2:     **return** *null*
3: $(M', type) \leftarrow FLMM(M, \alpha, T_p, T_e)$
4: **if** *type* is *microlensing* **then**
5:     $d \leftarrow DTW(M', D_m)$
6: **if** *type* is *flarestar* **then**
7:     $d \leftarrow DTW(M', D_f)$
8: **if** $d < d_{max}$ **then**
9:     **return** *type*
10: **return** *null*

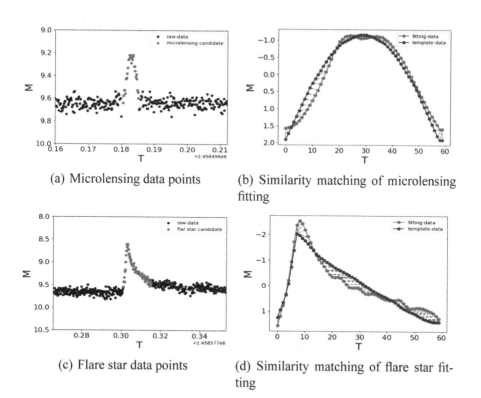

(a) Microlensing data points

(b) Similarity matching of microlensing fitting

(c) Flare star data points

(d) Similarity matching of flare star fitting

Fig. 7. Similar matches using template data

Through Raw Data Detection. The influence of geographical environment and bad weather conditions on the original data collection can lead to the phenomenon of original data faulting as shown in Fig. 8. After the data preprocessing, the fault phenomenon will be gradually stabilized and become smooth, which shows the characteristics of the light event as shown in the red curve in Fig. 8. In this case, the abnormal data can easily be mistaken for normal light event.

In order to remove the false light event formed from the conversion of the original data, we substitute the original data corresponding to the candidates obtained above into the algorithm for detection. We delete the data that does not meet the conditions, thereby improving the accuracy of light event recognition.

5   Experimental Evaluation

The evaluation is conducted in a desktop PC (Intel(R) Core(TM) i7-4790CPU, 3.6 GHz, 8 GB memory, 1 TB disk) running Ubuntu 14.04 (64 bits, kernel version 4.8.2-19). The program is programmed in the development tool PyCharm (64 bit, version 2020.1.2 (Professional Edition)) through python (version 3.7.7). The Python library files used are mainly PYWT and UCRDTW. The paper uses real data collected by GWAC. The average sampling frequency of the data is 15 s, and the data set contains about 0.93 million pieces of data.

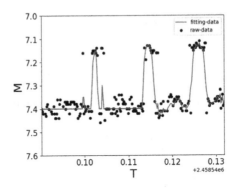

Fig. 8. Abnormal data is identified as a light event

We test the operating efficiency and the accuracy of detection results on different data sets by using FLMM, DTW and FLMM-DTW. Figure 9(a) shows the comparison of operating efficiency on different data sizes. The operating efficiency of the three algorithms has no obvious running speed problem when running on 10,000 data. When the data set reaches a million pieces, the operating efficiency of FLMM is the highest, followed by FLMM-DTW. The use of DTW requires all data to be matched, which shows the lowest operating efficiency. Figure 9(b) shows the number statistics of the detection results of the three algorithms. In the test of the whole data set, the results detected by DTW reach up to 20,000 pieces of data. Due to the rarity of light events, there must be a large number of abnormal data among these massive data. The data volume of FLMM-DTW is about two orders of magnitude lower. Figure 9(c) shows the accuracy rates of the three algorithms in detecting results on various data scales. In the case of less data, the difference between DTW and FLMM-DTW is small. As the data scales to the millions, there is a significant difference in the accuracy of the results. The accuracy of FLMM-DTW is almost two orders of magnitude higher than DTW.

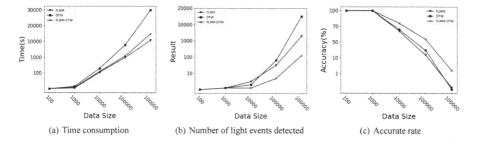

(a) Time consumption   (b) Number of light events detected   (c) Accurate rate

Fig. 9. Comparison of three algorithms

After comparing the operating efficiency and accuracy of the above three algorithms, we test the algorithm with the best performance using the original data and the pre-processed data. Figure 10(a) shows the same algorithm to test the efficiency of different data sets. As expected, running with just preprocessed data is certainly faster than running with two data sets. Figure 10(b) shows the accuracy rate of the same algorithm using different data sets to detect the results. The accuracy rate is improved with the use of raw data and pre-processed

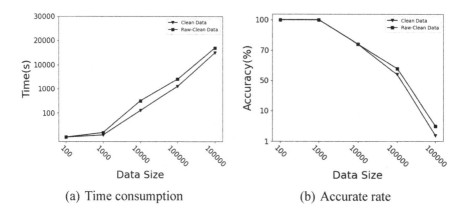

(a) Time consumption   (b) Accurate rate

Fig. 10. Comparison of different data sets

data. Although the two data sets have a doubled relation, we only perform algorithmic calculations on the original data of the detect results, so the running time does not increase too much. Improving the accuracy of test results within an acceptable time cost is very feasible range.

6   Conclusion

We propose a fast location method to detect light events of celestial bodies. Our method is nearly three times faster than DTW matching, and the accuracy of the detection results is improved by two orders of magnitude.

Acknowledgement. This work is supported by NSFC under grants 61972198, Natural Science Foundation of Jiangsu Province of China under grants BK20191273.

References

1. Brillinger, D.R.: Time series - data analysis and theory. In: Classics in Applied Mathematics, vol. 36. SIAM (2001)
2. Cai, Q., Xie, Z., Chen, G., Jagadish, H.V., Ooi, B.C., Zhang, M.: Effective temporal dependence discovery in time series data. Proc. VLDB Endow. **11**(8), 893–905 (2018)
3. Chiang, Y., Doan, A., Naughton, J.F.: Modeling entity evolution for temporal record matching. In: ACM SIGMOD, pp. 1175–1186 (2014)
4. Dignös, A., Böhlen, M.H., Gamper, J.: Temporal alignment. In: ACM SIGMOD, pp. 433–444 (2012)
5. Feng, T., Du, Z., Sun, Y., Wei, J., Bi, J., Liu, J.: Real-time anomaly detection of short-time-scale GWAC survey light curves, pp. 224–231. IEEE (2017)
6. Gao, J., Agarwal, P.K., Yang, J.: Durable top-k queries on temporal data. Proc. VLDB Endow. **11**(13), 2223–2235 (2018)
7. Li, F., Yi, K., Le, W.: Top-k queries on temporal data. VLDB J. **19**(5), 715–733 (2010)
8. Paczynski, B.: Gravitational microlensing by the galactic halo. Astrophysical J. **304**(1) (1997)
9. Schneider, P., Ehlers, J., Falco, E.: Gravitational Lenses. Springer (1992). https://doi.org/10.1007/978-3-662-03758-4
10. Smith, C.J., Villanueva, G.L., Suissa, G.: Imagining exoplanets: visualizing faraway worlds using global climate models. In: ACM SIGGRAPH, pp. 20:1–20:2 (2020)
11. Song, S., Cao, Y., Wang, J.: Cleaning timestamps with temporal constraints. Proc. VLDB Endow. **9**(10), 708–719 (2016)
12. Song, S., Zhang, A., Wang, J., Yu, P.S.: SCREEN: stream data cleaning under speed constraints. In: ACM SIGMOD, pp. 827–841 (2015)
13. Wan, M., Wu, C., Zhang, Y., Xu, Y., Wei, J.: A pre-research on GWAC massive catalog data storage and processing system. Astron. Res. Technol. **13**(3), 373–381 (2016)
14. Xu, J., Liang, J.: A system for querying and displaying typed intervals. In: Gertz, M., et al. (eds.) SSTD 2017. LNCS, vol. 10411, pp. 440–445. Springer, Cham (2017). https://doi.org/10.1007/978-3-319-64367-0_31

15. Xu, J., Lu, H.: Efficiently answer top-k queries on typed intervals. Inf. Syst. **71**, 164–181 (2017)
16. Yi, H., Ouyang, P., Yu, T., Zhang, T.: An algorithm for Morlet wavelet transform based on generalized discrete Fourier transform. Int. J. Wavelets Multiresolution Inf. Process. **17**(5), 1950030 (2019)
17. Zhang, D., Gunopulos, D., Tsotras, V.J., Seeger, B.: Temporal aggregation over data streams using multiple granularities. In: Jensen, C.S., et al. (eds.) EDBT 2002. LNCS, vol. 2287, pp. 646–663. Springer, Heidelberg (2002). https://doi.org/10.1007/3-540-45876-X_40
18. Zhang, D., Markowetz, A., Tsotras, V.J., Gunopulos, D., Seeger, B.: On computing temporal aggregates with range predicates. ACM Trans. Database Syst. **33**(2), 12:1–12:39 (2008)

Efficiently Update Disk-Resident Interval Tree

Jianqiu Xu$^{(\boxtimes)}$ and Jianhua Wei

Nanjing University of Aeronautics and Astronautics, Nanjing, China
{jianqiu,jianhua}@nuaa.edu.cn

Abstract. Supporting frequent update throughput is an essential issue in applications that involve monitoring and querying continuous variables. We present an I/O optimal method to efficiently update the disk-resident interval tree for a set of new intervals. The idea is to partition the input data into two parts and bulk-applies each part into the structure. Meanwhile, the tree balance is preserved. We introduce our proposal and develop alternative methods. To verify the performance, an experimental evaluation is conducted. The results demonstrate that our method achieves three orders of magnitude better performance than individual updates and 1.5–3 times faster than the *drop-and-rebuild* method for updating 0.5 million intervals on the historical data containing 10 million intervals.

Keywords: Interval tree · Update · Bulk load

1 Introduction

A number of emerging applications require the database system to manage not only the historical data but also the incoming data. A significant task is to synchronize index structures in order to be consistent with the underlying data space. Further, users are allowed to efficiently query and analyze these data.

In this paper, we study efficiently updating the disk-resident interval tree for a large amount of new intervals. The interval tree [9,14] has been widely used in spatial and temporal databases to manage intervals representing axis-parallel line segments. For example, intervals occur as line segments on a space-filling curve [5] and define transaction time and valid time ranges [6].

An interval tree is built on intervals, each of which consists start and end points. In principle, an interval tree is a binary tree that serves as the primary structure. Each node maintains a value called the split point and two lists of sorted intervals that intersect the split point, called the secondary structure. Intervals smaller and larger than the split point are stored in the left and right subtrees, respectively. The creating procedure is unique in the sense that all input data are known in advance. This is because one needs to determine *min* and *max* endpoints to partition the data. For many commonly used index structures such as B-trees and R-trees, the creating algorithm does not have to know the

X. Meng et al. (Eds.): SpatialDI 2020, LNCS 12567, pp. 198–207, 2021.
https://doi.org/10.1007/978-3-030-69873-7_14

data range. In the literature, a large amount of efforts have been made on bulk loading index structures [7,8], i.e., creating an index from scratch. The bulk loading procedure is in contrast to inserting the input data into the index one by one. We do not deal with bulk loading an interval tree, but the main concern is to effectively insert a batch of new intervals into an interval tree built already in a batch update scenario. Figure 1 shows an interval tree built on intervals $\{o_1, o_2, ..., o_8\}$. Consider three new arrival intervals: o_1' , o_2' and o_3' . Updating the structure imposes two challenges.

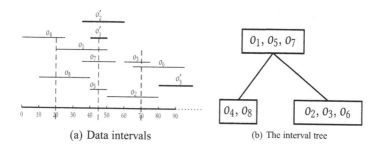

(a) Data intervals   (b) The interval tree

Fig. 1. A running example

The I/O Efficiency. For databases supporting update-intensive workloads, the efficiency of disk I/O is an effective means for improving performance and scalability. Given a set of n intervals, inserting a new interval requires $O(\log n)$ to find the node and incurs $O(\log_b n)$ I/Os [3], in which b is the block size. Iteratively applying this procedure is slow and costly for a large amount of update data. Given a set of n new intervals, the structure is traversed $O(n)$ times, incurring high CPU and I/O costs. One tree node may be accessed several times for updating different intervals. In the example, o_1' and o_2' will be inserted into the same node and two node accesses are required. In fact, one time read/write operation is sufficient. One could also drop and rebuild the index, but repeating this operation for each update is extremely expensive. If the update only includes a few intervals, one still has to rebuild the whole index. Therefore, this method is limited in scope.

Balance. The new interval may not intersect any existing node, e.g., o_3'. A new node is created to hold the interval. When inserting a node into the structure, the tree balance condition may not hold. Therefore, a rotation operation is performed. When nodes change their positions in the tree, intervals may be moved because the following criteria must be met: each interval is located in the first intersecting node from top to bottom level.

Motivated by this, we propose an efficient algorithm to update the structure by bulk load and perform the rebalancing if required. Special-purpose bulk loading algorithms often perform significantly better than the one using repeated insertion [2]. Bulk operations are of interest for any kind of indexes, but most of the available methods are presented in the context of B-tree [10] and R-trees related structures [2]. Little attention has been paid to other structures. To the best of our knowledge, no method in the literature has addressed updating the interval tree by bulk load. The paper [3] develops an I/O-optimal method to update an interval in $O(\log_b n)$ I/Os amortized. Repeating this operation takes $O(n \cdot \log_b n)$ I/Os to update n intervals. The balance problem is not considered because they assume that all new intervals intersect existing nodes.

Our method is to partition new intervals into two parts: one contains intervals located in the current data space such as o_1' and o_2', and the other contains intervals out of the current data space, for example, o_3'. For the first part, we traverse the tree once to determine the nodes in which new intervals will be located. Intervals in the second part will incur creating new nodes. Instead of inserting new intervals one by one, we build a second interval tree and insert each node of the second interval tree into the existing structure. Each node to be inserted may contain several intervals and the time of separate write I/Os is reduced because the number of nodes is usually much smaller than the number of intervals.

We develop the bulk load algorithm and two alternative methods: individual updates and the drop-and-rebuild method. Using large real and synthetic datasets, we conduct experiments in a prototype database system to evaluate the update efficiency and the update impact on query performance. The results demonstrate that (i) using 10 millions intervals as the historical data and 0.5 million new intervals, our method achieves three orders of magnitude better performance than individual updates and outperforms the drop- and-rebuild method by a factor of 1.5; (ii) updating the structure by bulk load does not influence the query performance.

The rest of the paper is organized as follows. The related work is discussed in Sect. 2. We review the interval tree and present alternative methods in Sect. 3. Our bulk load algorithm is introduced in Sect. 4. The evaluation is reported in Sect. 5. We conclude the paper in Sect. 6.

2   Related Work

Updating Indexes. In the literature, there is a lot of related work on updating indexes including traditional index structures and spatial indexes. High updates for B-tree are discussed in [10]. The paper [12] investigates efficient update algorithms for structure indexes including two kinds of updates: the addition of a new file to the database and a small incremental change. Based on the notion of graph bi-similarity, three instances of structure indexes are processed: 1-Index, F&B-Index and A(k)-index. The proposed algorithms are an order of magnitude faster than dropping and rebuilding the index. Compared with our work, there are fundamental differences. We focus on adding new intervals into the database instead of modifying existing data intervals (no rebalance). This looks like the addition of a new file to the database. However, that work views the data as

XML files and the addition of a new file corresponds to the addition of a sub-graph under the root. Intervals are not defined by graph models. We deal with the interval tree which is different from 1-Index, F&B-Index and A(k)-index.

Efficiently updating R-trees has been extensively investigated. A batch of updates on R-tree is studied in [15] in which the rectangles to be inserted are sorted and then packed into blocks. These blocks are inserted one at a time using standard insertion algorithms. *L. Arge* et al. propose a complete bulk operations on dynamic R-trees including loading, insertion, deletion and queries [3]. *C.S. Jensen* et al. exploit the buffering of updating operations in main memory as well as the grouping of operations to reduce disk I/Os [4]. An R-tree based index structure with update strategies is proposed in [17] to support high update throughput. Each update means a delete-insert operation. Other methods include the bottom-up approach [13] and using memos [16]. All those methods attach buffers to tree nodes and bulk-apply the buffered updates to the index if the buffer fills up. Clustered rectangles (groups) in the buffer are inserted one by one. Since one operation updates a group of objects, the I/O cost is minimized.

We update an interval tree which is not an R-tree related structure. This results in different update operations. We perform an rotation operation to keep the interval tree balanced, but this does not occur for R-tree. We do not attach buffers to all nodes but maintain one for the root node. If buffers are attached to all nodes, the time to flush is not easy to determine because the input data may be not uniformly distributed in tree nodes.

3   Preliminary

3.1   Interval Tree

Let $O = \{[s, e] | s, e \in \mathcal{R}, s < e\}$ be a set of intervals, each of which contains start and end points whose universe is defined in a real domain. To build an interval tree, intervals are sorted in ascending order according to start and end points. The structure is built recursively from the root down following the procedure: Step 1, a split point is computed, denoted by p. Step 2, we use p to divide the interval set into three parts: (i) intervals fully to the left of p; (ii) intervals containing p; (iii) intervals fully to the right of p. The split point should be picked in such a way that the tree is relatively balanced, usually the median point. Step 3, a node is created to hold part (ii) and two child pointers lp and rp are defined for nodes maintaining (i) and (iii), respectively. We repeat Steps 1–3 for (i) and (iii) until no interval is left. The *primary structure* is a balanced binary tree and each internal node is associated with two lists of intervals that form the *secondary* structure, mapping to records in the disk.

3.2   Alternative Methods

Given a set of new intervals, there are two alternative methods to update the interval tree. One is to drop and rebuild the index, called *DropRebuild*. This method is useful and efficient for small datasets but not large datasets. On one hand, rebuilding the complete index is a costly procedure, especially when the update data comes continuously. On the other hand, new intervals may only update a few nodes and hence it is not necessary to rebuild the structure.

The other approach is to iteratively insert each new interval into the structure, called *Baseline*. The inserting procedure incurs updating a node or creating a new node. The former happens if the interval intersects the split point of a node. Then, left and right lists in the node are updated. The latter occurs when the new interval does not intersect any node. In this case, a node is created to hold the new interval. When inserting a node into the structure, a rotation may be performed to keep the tree balanced.

According to the tree property, an interval is stored in the first node whose split point intersects the interval by following the top-down traversal. The rotation changes the parent-child relationship and therefore intervals may be moved among rotated nodes. An example is shown in Fig. 2. After inserting o'_4, we perform a single rotation and have the node containing o'_3 as the root node of the subtree. o_6 will be moved to the new root node because o_6 contains the split point.

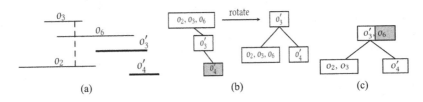

Fig. 2. Perform the rotation and move intervals

4   Insert by Bulkload

4.1   Partition the Data

Let $Min = min(o.s)$ and $Max = max(o.e)$ $(o \in O)$ be the minimum and maximum endpoints. We assume that applications cause an ongoing expansion of the data space, i.e., new intervals only update Max. Given a new interval o', there are three relationships between $[o'.s, o'.e]$ and $[Min, Max]$: case (i): $Max \geq o'.e$; case (ii): $Max < o'.s$; case (iii): $o'.s < Max < o'.e$, as shown in Fig. 3.

If o' falls in case (iii), it is split into $\{o'_{in}, o'_{out}\}$ in which $o'_{in} = [o.s', Max]$ and $o'_{out} = [Max, o.e']$. The first part goes to Case (i) and the second part goes to Case (ii). The two intervals still have the original id pointing to the complete interval and are flagged in order to distinguish from non-split intervals.

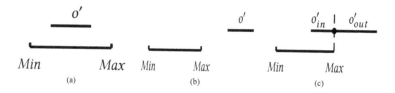

Fig. 3. Relationship between o' and [Min, Max]

4.2 The Algorithm

Let T denote an interval tree. We give the overall algorithm in Algorithm 1. After the partition, intervals in O'_{in} and O'_{out} are processed individually by calling two subroutines. Updated nodes are invoked once by the flush algorithm to process all updates.

Algorithm 1. $BulkUpdate(T, O')$

1: *partition O' into $\{O'_{in}, O'_{out}\}$*
2: *Insert_In(T, $T.RootId()$, O'_{in})*
3: *Insert_In(T, $T.RootId()$, O'_{out})*

Process O'_{in}. We insert new intervals into the structure by performing one time traversal. Starting from the root node, for each accessed node we partition new intervals into three parts: (i) O_l, intervals will be stored in the left subtree; (ii) O_r, intervals will be stored in the right subtree; and (iii) O_m, intervals stored in this node. We visit the left child for O_l and the right child for O_r. Left and right lists of this node are updated by O_m. Each node is accessed and updated only once. The algorithm is given in Algorithm 2. We insert $O'_{in} = \{o'_1, o'_2, o'_5, o'_{3a}\}$ into the structure, and have $O_l = \{o'_5\}$, $O_m = \{o'_1, o'_2\}$ and $O_r = \{o'_{3a}\}$ after accessing the root node.

Rebalance. If a new interval falls in between two tree nodes or does not intersect any existing node such as o'_5, a new node is created and will be inserted into the existing structure. If there is a set of new intervals, an interval tree will be created. After that, we insert each node of the new tree into the existing structure. Inserting nodes will eventually cause rebalancing and moving intervals.

Process O'_{out}. We build an interval tree on O'_{out}, denoted by T', and then merge each node in T' into T. Compared to iteratively inserting each new interval into T, inserting each node in T' into T greatly reduces the time of accessing the interval tree because each node in T' may contain a set of intervals. During the procedure, we may perform the rotation and move intervals.

Correctness Discussion. First of all, we guarantee the integrity of the data when partitioning the data, that is, there will be no data omission and duplication. Because when we deal with the data that exceeds and intersects the original data

Algorithm 2. $Insert\_In(T, \text{id}, O'_{in})$

1: $O_l \leftarrow \varnothing, O_r \leftarrow \varnothing, O_m \leftarrow \varnothing$ $N \leftarrow \text{GetNode}(T, id)$
2: **for all** $o' \in O'_{in}$ **do**
3:     **if** o' contains $N.p$ **then**
4:         $O_m \leftarrow o'$
5:     **else**
6:         **if** $N.p < o'$ **then** $O_r \leftarrow o'$
7:         **if** $N.p > o'$ **then** $O_l \leftarrow o'$
8: *update left and rights lists in N by O_m*
9: **if** $N.lp > 0$ **then**
10:     $Insert\_In(T, N.lp, O_l)$
11: **else**
12:     *create a subtree on O_l and insert into $N.lp$*
13: **if** $N.rp > 0$ **then**
14:     $Insert\_In(T, N.rp, O_r)$
15: **else**
16:     *create a subtree on O_r and insert into $N.rp$*

Algorithm 3. $Insert\_Out(T, O'_{out})$

1: create an interval tree T' on O'_{out}
2: **for all** $N' \in T'$ **do**
3:     *insert N' into T*
4:     //Implements the merging of nodes created for the intervals in O'_{out} with the existing interval tree
5:     **if** T is not balanced **then**
6:         *perform the rotation and move intervals*

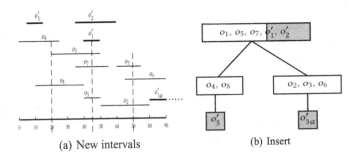

(a) New intervals   (b) Insert

Fig. 4. Example of inserting O'_{in}

range, we will divide such data into two parts, o'_{in} and o'_{out}. The two intervals still have the original id pointing to the complete interval and are flagged in order to distinguish from non-split intervals. In addition, during the batch update process, we access the existing interval tree from top to bottom. The processed data is inserted according to the interval tree construction rules and the balance of the tree is adjusted, thereby ensuring the correctness of the tree.

5   Experimental Evaluation

The evaluation is conducted in a standard PC (Intel(R) Core(TM) i7-4770CPU, 3.4 GHz, 8 GB memory, 2 TB disk) running Ubuntu 14.04 (64 bits, kernel version 4.8.2-19). We develop the index structure and algorithms in C/C++ and integrate the implementation into an extensible database system SECONDO [11].

Both synthetic and real datasets are used, as reported in Table 1. The synthetic dataset is generated as follows. The start point of an interval is randomly chosen within the domain [Min, Max], and the length is a stochastic value $\in$ [1, 1000]. The real dataset is from a data company DataTang [1]. Taxi stores GPS records of 30,000 Beijing taxis in two days. Each interval represents the span between two continuous GPS records with the average length 47. The overall data includes 22.29 million intervals.

Table 1. Dataset statistics

	Historical Data		Update Data		
	$|O|$(million)	[Min, Max]	$|O'|$		[Min, Max]
Syn	{0.1, 0.5, 1,5, **10**}	[1, 100000]	{5, 50, 500, **5000**, 50000, 500000}		[1, 120000]
Taxi	20		[1, 173459]	{5, 50, 500, **5000**, 50000, 500000}	[1, 173459]

Update Efficiency. We perform the evaluation by scaling the number of update intervals. For the synthetic dataset, 10 million intervals are considered as historical data and update intervals are randomly generated over the domain. For the real dataset, 20 million intervals are considered as historical data and update intervals are randomly selected from the rest. The results are reported in Figs. 5(a) and 5(b), demonstrating that (i) the performance difference between *Baseline* (individual updates) and bulk load is marginal for a small number of updates; (ii) the drop-and-rebuild method is not sensitive to the number of update intervals because normally $|O'| \ll |O|$; and (iii) using large datasets our

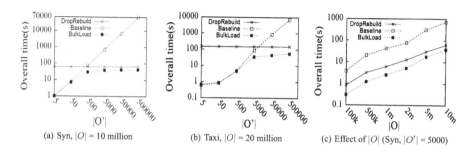

(a) Syn, $|O|$ = 10 million   (b) Taxi, $|O|$ = 20 million   (c) Effect of $|O|$ (Syn, $|O'|$ = 5000)

Fig. 5. Scale $|O'|$ and $|O|$

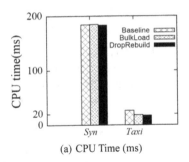
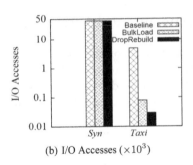

(a) CPU Time (ms)   (b) I/O Accesses ($\times 10^3$)

Fig. 6. Query cost

method is more than an order of magnitude faster than Basline and outperforms the drop-and-rebuild method by a factor of 1.5–3. The effect of historical data size is also tested. Using the synthetic dataset, we consider 5000 update intervals on different numbers of historical intervals. The results are averaged over 20 runs, as shown in Fig. 5(c). When the historical data size is enlarged, the costs of all methods increase but our approach is 2 times faster than *DropRebuild* and more than an order of magnitude faster than *Baseline*.

Query Efficiency. We test whether the query efficiency is affected by bulk load updating. We update $|O'| = 500,000$ intervals and then test the query efficiency. Intersecting queries are evaluated. Each interval is assigned a weight and top-k ($k = 10$) intervals are returned. CPU time and I/O accesses are defined as performance metrics. The results are averaged over 20 runs. We report the query cost in Fig. 6, showing that updating the structure will not influence the query performance.

6   Conclusion

We develop an efficient algorithm to update an interval tree by bulk load. Our method substantially outperforms individual updates and is also 1.5–3 times faster than the drop-and-rebuild method.

Acknowledgment. This work is supported by NSFC under grants 61972198, Natural Science Foundation of Jiangsu Province of China under grants BK20191273 and National Key Research and Development Plan of China (2018YFB1003902).

References

1. http://factory.datatang.com/en/ (2018)
2. Arge, L., Hinrichs, K.H., Vahrenhold, J., Vitter, J.S.: Efficient bulk operations on dynamic r-trees. Algorithmica **33**(1), 104–128 (2002)
3. Arge, L., Vitter, J.S.: Optimal external memory interval management. SIAM J. Comput. **32**(6), 1488–1508 (2003)

4. Biveinis, L., Saltenis, S., Jensen, C.S.: Main-memory operation buffering for efficient r-tree update. In: VLDB, pp. 591–602 (2007)
5. Böhm, C., Klump, G., Kriegel, H.-P.: Xz-ordering: A space-filling curve for objects with spatial extension. In: 6th International Symposium, Advances in Spatial Databases, pp. 75–90 (1999)
6. Bozkaya, T., Ozsoyoglu, M.: Indexing valid time intervals. In: Quirchmayr, G., Schweighofer, E., Bench-Capon, T.J.M. (eds.) DEXA 1998. LNCS, vol. 1460, pp. 541–550. Springer, Heidelberg (1998). https://doi.org/10.1007/BFb0054512
7. den Bercken, J.V., Seeger, B.: An evaluation of generic bulk loading techniques. In: VLDB, pp. 461–470 (2001)
8. den Bercken, J.V., Seeger, B., Widmayer, P.: A generic approach to bulk loading multidimensional index structures. In: VLDB, pp. 406–415 (1997)
9. Edelsbrunner, H.: Dynamic data structures for orthogonal intersection queries. Tech. Univ. Graz, Austria, Technical report (1980)
10. Graefe, G.: B-tree indexes for high update rates. SIGMOD Rec. **35**(1), 39–44 (2006)
11. Güting, R.H., Behr, T., Düntgen, C.: SECONDO: a platform for moving objects database research and for publishing and integrating research implementations. IEEE Data Eng. Bull. **33**(2), 56–63 (2010)
12. Kaushik, R., Bohannon, P., Naughton, J.F., Shenoy, P.: Updates for structure indexes. In VLDB, pp. 239–250 (2002)
13. Lee, M., Hsu, W., Jensen, C.S., Cui, B., Teo, K.L.: Supporting frequent updates in R-trees: a bottom-up approach. In: VLDB, pp. 608–619 (2003)
14. Overmars, M., Berg, M., Kreveld, M., Schwarzkopf, O.: Computational Geometry. Section 10.1: Interval Trees. Springer, Heidelberg (2000). https://doi.org/10.1007/978-3-662-04245-8
15. Khalil, M., Kamel, I., Kouramajian, V.: Bulk insertion in dynamic R-trees. In: SDH, pp. 31–42 (1996)
16. Silva, Y.N., Xiong, X., Aref, W.G.: The RUM-tree: supporting frequent updates in R-trees using memos. VLDB J. **18**(3), 719–738 (2009)
17. Song, M., Choo, H., Kim, W.: Spatial indexing for massively update intensive applications. Inf. Sci. **203**, 1–23 (2012)

Performance Ranking Based on Bézier Ranking Principal Curve

Wei Jin[1,2(✉)], Danhuai Guo[3,4], Li-kun Zhao[5], and Ji-Chao Li[1,2]

[1] Beijing Academy of Science and Technology, Beijing 100094, China
jinwei201002@163.com
[2] Beijing Institute of New Technology Applications, Beijing 100094, China
[3] Computer Network Information Center, Chinese Academy of Sciences, Beijing, China
[4] University of Chinese Academy of Sciences, Beijing, China
[5] School of Civil Engineering, North China University of Technology, Beijing 100144, China

Abstract. As for the problem of subjectivity and lack of scientificity in performance evaluation, the paper introduces a RPC model with cube Bézier curve being principal curve, and designs the learning algorithm of Bézier curve. First, evaluation indicators are chosen using Pearson correlation coefficient, RPC two-dimensional projection is conducted on relevant data and the ranking result of multiple indicator observation data is obtained. The experiment is done with the performance data of 12345 hotline telephone operators in one month of 2019 in one district of Beijing, and the ranking result is in line with the popular perception. The experimental result proves that as for the ranking in performance evaluation, RPC method is more scientific and interpretable than the method of weighted average and Elmap model.

Keywords: Unsupervised ranking · Principal curves · Multiple indicators · Performance evaluation · Bézier curve

1 Introduction

In the area of government administration, the government needs to evaluate the staff's performance. Through establishing a scientific evaluation mechanism, the government conducts the work evaluation of the staff, which is conducive to government's optimal management and improving staff's work initiative.

From the perspective of machine learning, performance evaluation of the staff is unsupervised ranking. Usually, the evaluated staff get the comprehensive scores according to the ranking rules of multiple indicators and then they are ranked according to their comprehensive scores. Therefore, when the ranking rules are made, observation data of all the evaluated staff in various indicators should be studied comprehensively so that more rigorous ranking can be obtained. With the rapid development of computer science, more and more machine learning methods are applied to comprehensive evaluation. By means of unsupervised learning methods with which multi-indicator observation data of the research objects are ranked, the percentage of subjective judgment in evaluating the

X. Meng et al. (Eds.): SpatialDI 2020, LNCS 12567, pp. 208–217, 2021.
https://doi.org/10.1007/978-3-030-69873-7_15

key indicators can be reduced. In model construction learning, there are two challenges in unsupervised learning methods. The first one is how to guarantee the reasonability of ranking models when there is no real ranking result. The second one is how to measure the degree of importance of ranking indicator to ranking.

As for the ranking in specific application domain, how to combine the domain knowledge with ranking models and to rank the objects reasonably is the hot issue attracting the attention of many government regulators and scholars, for example, the choice of domain knowledge, of models, and of indicators, and the control of algorithm complexity. Hu et al. proposed a machine learning model frame driven jointly by "data" and "knowledge" [31, 32]. In literature [37, 38], when ordinal classification was conducted with Decision Tree, monotonous priori knowledge was applied to learning algorithm, which improved the ordinal classification accuracy of Decision Tree. Although Klementiev et al. proposed the unsupervised fusion method based on the ranking results of experts in this field [40], invariance of scale translation and monotonicity were not discussed. The application of domain knowledge to unsupervised image segmentation evaluation and to clustering evaluation provides ranking domain knowledge embedded in unsupervised sorting models with a method of domain knowledge embeddedbness model.

For the performance evaluation of telephone operators of hot line 12345 in some district of Beijing in 2019, with the observation data of ranked objects in multiple evaluating indicators, in the condition that there is no real ranking order, the paper embeds the priori knowledge of ranking issue in the ranking models in the parameterized way and discusses the degree of importance of each evaluation indicator to models with Spearman's rank correlation coefficient. This ranking model can describe smoothly linear and nonlinear distributed structure of data and its geometric significance of parameters is explicit. It avoids the issue of how to choose complicated models.

The structure of the paper is as follows. Section 2 summarizes related work. Section 3 describes ranking principal curve of Bézier curve construction and learning algorithm of ranking principal curve. Section 4 describes the application of ranking principal curve model to performance evaluation of telephone operators of hot line 12345 and lays special stress on analyzing the ranking result. Section 5 summarizes the whole paper and puts forward the future research direction.

2   Related Works

As for multiple-indicator observation data, the unsupervised ranking methods can be divided into two types: rank aggregation and structure ranking. Probabilistic fusion models blended multiple sequences with objective function of maximum likelihood [9, 10], such as, Multinomial Preference Model, MPM for short [11, 12]. However, almost all the probabilistic fusion models are implicit ranking models, and there is no explicit expression for ranking rules. Therefore, it is impossible to make sure whether the model score can keep the same order with the real order in the original data space (that is, order-preserving or strict monotonicity [15]). Moreover, as complexity of models of probabilistic fusion ranking models is unknown, there lacks fair basis in comparison with other ranking models.

Structure ranking method is to rank research objects according to the data structure. The most intuitive way is to find a data skeleton as "ranking coordinate" to rank the research objects, and the representative method is Principal Component Analysis (PCA). However, PCA model is very simple. It can only be applied to oval data distribution, and not to crescent data distribution. Gorban et al. proposed an Elastic Map (Elmap for short) promoted by the idea of Manifold Ranking [18, 19] and it solved this problem. Elmap can describe nonlinear data skeleton, but it has no specific ranking rules expression. The complexity of models is hard to determine and, it cannot guarantee the satisfaction of order-preserving. Similarly, the methods of other principal curve models [22, 23] are also used for ranking. Moreover, although structure ranking method can provide a ranking frame, there still lacks effective theory of guarantee and explicit ranking evaluation for the reasonability of ranking results, which restricts the development of unsupervised ranking models. Li Chunguo proposed to promote ranking to nonlinear space with parameterized ranking principal curve model of cubic Bézier curve, but as for the multiple-indicator ranking issue, the algorithm cannot solve the problem of which indicators should be chosen for ranking. Therefore, the paper proposes the application of ranking principal curve model based on cubic Bézier curve and Spearman rank correlation coefficient to the performance evaluation of telephone operators of hot line 12345, which can show the scientificity and reasonability of this model and can solve the problem of subjectivity of manual ranking.

3   Proposed Method

3.1   The Definition of Ranking Principal Curve Model

The definition of ranking principal curve is as follows.

$$\min J(\theta) = \sum_{i=1}^{n} \|x_i - f(s_i, \theta)\|^2 \tag{1}$$

In d-dimensional space, if one-dimensional space curve $f(s, \theta)$ satisfies $f(s, \theta)$ is a strictly monotonous principal curve;

$|\theta|$ is definite;

$$\left(\frac{\partial f}{\partial s}\right)^T (x_i - f(s, \theta))|_{s=s_i} = 0, i = 1, 2, \cdots, n \tag{2}$$

Where $|\theta|$ represents the cardinality of θ, $f(s, \theta)$ is Ranking Principal Curve, RPC for short.

With observation data $X = \{x_1, x_2, \cdots, x_n\}$ of a group of ranking objects in d evaluation indicator, the ranking process of research objects according to principal curve is an issue of constrained nonlinear optimization. According to the definition of general constraint (that is, $g(x, \theta) \propto R_i(f)$ defined in [25], where symbol "" shows the coupling relationship between the two), 5 ranking meta criteria about ranking rules in ranking issues are constraint conditions of nonlinear issues, which makes the learned principal curve satisfies monotonous constraint and makes it ranking principal curve. The general nonlinear planning model corresponding to ranking issue based on principal curve is:

$$f(s, \theta) \propto R_j(f), j = 1, 2, 3, 4, 5 \tag{3}$$

Where formula (2) looks for points $f(s, \theta)$ in principal curve, which makes the vertical distance $x_i - f(s_i, \theta)^2$ from observation vector xi to principal curve shortest (or the reconstruction error is minimum); $R_j(f)(j = 1, 2, 3, 4, 5)$ is constraint condition corresponding to ranking meta criteria.

3.2 Ranking Principal Curve Constructed by Bézier Curve

If the constraint of control point of Bézier curve is added, cube Bézier curve can describe four basic monotonous curve in two-dimensional curve. The paper adopts cube Bézier curve to construct ranking principal curve, and the formula of constructing model is

$$f(s) = \sum_{r=0}^{3} B_r^3(s)p_r = PMz, s \in [0, 1] \tag{4}$$

Four control points of Bézier curve are parameters of ranking principal curve, and the number of parameters is $|\theta| = 4d$. Specially, cube Bézier curve satisfies the other four meta criteria of evaluation function.

3.3 Learning Algorithm of Ranking Principal Curve

With measurement matrix $X = (x_1, x_2, \cdots, x_n)$ of research objects in multiple indicators, ranking principal curve is constructed using cube Bézier curve, approximation performance of observation data skeleton from principal curve is determined by control point, and P_0, P_1, P_2 and P_3 are parameters to be optimized of ranking principal curve. According to optimization model of ranking principal curve, constrained nonlinear optimization issue corresponding to cube Bézier curve can be obtained as follows:

$$\min J(P, s) = \sum_{i=1}^{n} \|x_i - PMz_i\|^2 \tag{5}$$

$$s.t. \left(\frac{\partial PMz}{\partial s}\right)^T (x_i - PMz)|_{s=s_i} = 0 \tag{6}$$

Where since cube Bézier curve itself satisfies other ranking meta criteria constraint except strict monotonicity, constraint formula (7) and formula (8) are added to control point of cube Bézier curve, which satisfied the definition requirement of ranking principal curve.

$$P_0 = \frac{1}{2}(1 - \alpha) \quad P_3 = \frac{1}{2}(1 + \alpha) \tag{7}$$

$$p_1, p_2 \in (0, 1)^d \tag{8}$$

Learning algorithm of ranking principal curve summarizes the whole iterative process of optimization algorithm solution.

$$s = (s_1, s_2, \cdots, s_n), z_i\left(1, s_i, s_i^2, s_i^3\right)^T$$

$$s_i \in [0, 1], i = 1, 2, \cdots, n. \tag{9}$$

In step 1, normalized pretreatment conducts is conducted to data.

In step 2, before control point $P^{(0)}$ is initialized, monotonous sign α of data needs to be determined. Monotonous sign shows the monotonous relationship of evaluation indicator to evaluation score. If the user is not given, monotonous sign can learn from data. According to principal component analysis of data, the difference of sample point corresponding to the maximum principal component and minimum principal component is calculated, and the positive or negative of component of alpha is determined by the positive or negative of the difference. The position of the initialized endpoint is $p_0 = \frac{1}{2}(1 - \alpha)$ and $p_3 = \frac{1}{2}(1 + \alpha)$, two sample points are chosen randomly as control points. In the learning process, with constant adjustment of its position, control point P(t) can be in the hypercube and strict monotonicity is satisfied, so that Bézier curve becomes ranking principal curve. The stopping rule of ranking learning algorithm is that the value of objective function J(P, s) no longer decreases, that is, when iteration stops.

$$\Delta J = J\left(P^{(t)}, s^{(t)}\right) - J\left(P^{(t+1)}, s^{(t+1)}\right) < 0 \tag{10}$$

Therefore, $\left(P^{(t)}, s^{(t)}\right)$ obtained is the control point which makes J local minimum point and the corresponding evaluation score.

Algorithm 1. Learning algorithm of ranking principal curve

Input:

  X: observation data of all research objects in multiple indicators;
  ξ: very small positive number;

Output:

  P*: control point of ranking principal curve;
  s*: score of all research objects.
  1: Normalize observation data of each indicator to [0,1];
  2: Initialize control point $P^{(0)}$;
  3: while $\Delta J > \xi$ do
  4: Determine the score $s^{(t)}$ of each research object using gold split method;
  5: Calculate $P^{(t+1)}$ according to formula (4.39);
  6: adjust control point to make it in hypercube $(0,1)^d$;
  7: if $\Delta J < 0$ then
  8: stop iteration;
  9: end if
  10: end while

---

4   Experiment

The paper adopts RPC method, Elmap model and traditional weighting method for contrastive analysis. First, normalization processing, interpolation calculation and feature selection of the data are conducted. Then, RPC two-dimensional projection of data is generated. Finally, several ranking results are analyzed. The experimental environment is open source software Scilab (version 5.4.1) and matlabR2017 with macOS-10.15.4 operating system based on 8G memory.

4.1 Data Collection

The data of this experiment come from 12345 non-emergency hotline of grid service administration center in one district of Beijing. Performance evaluation data of 10 telephone operators from January to December of 2019 are recorded, including the number of cases filed, the number of cases filed due to damage to public facilities, the number of filed cases which are not settled that very day, the number of cases distributed for settlement, the number of cases which are not settled, the number of cases settled, the number of garbage sorting sites, frequency of special inspection, telephone traffic, working hours, etc. The data are system export data and weighted calculation data. The data type of all indicators is numeric one.

4.2 Preprocessing

First, the data is preprocessed. With formula (11), the observation data of research objects in multiple indicators are normalized to [0,1], without changing evaluation scores and ranking position of research objects. In formula (11), $\hat{X}$ is the vector after the normalization of observation vector X of each research object in multiple indicators, and Xmin and Xmax are minimum observed vector and maximum observed vector of multiple indicators respectively.

$$\hat{X} = \frac{X - X_{min}}{X_{max} - X_{min}} \tag{11}$$

Because there are only the data of 10 telephone operators, the data size is too small for ranking, which will increase the calculation error. Therefore, before the ranking of principal curves, interpolation calculation method is adopted to extend the performance evaluation data of 10 telephone operators to 100 pieces of evaluation data, which makes data more evenly distributed and reduces the error. As the experimental data are multidimensional data, the method of oversampling is adopted to increase the data points. K-Nearest Neighbor is used in the process of oversampling. As for a data point to be predicted, k data points which have the shortest Euclidean distance to it are found, and their classification results are observed. The predicted result is the category which appears the most of k data points, so the prediction of the given data point category is achieved.

$$Euclidean\, distance(d) = \sqrt{(x_2 - x_1)^2 + (y_2 - y_1)^2} \tag{12}$$

Interpolation calculation finished, data size satisfying unsupervised ranking is obtained. However, due to many characteristic dimensions, the characteristics irrelevant to ranking will affect the ranking result. Therefore, it is necessary to choose characteristics and to eliminate redundant characteristics which do not affect ranking result in order to improve execution efficiency of unsupervised ranking. In the paper, Pearson correlation coefficient is adopted to choose the characteristics with the highest correlation as the ranking characteristics. The computational formula of Pearson correlation coefficient is:

$$\rho(X, Y) = \frac{E[(X - \mu_x)(Y - \mu_Y)]}{\sigma_X \sigma_Y} = \frac{[(X - \mu_x)(Y - \mu_Y)]}{\sqrt{\sum_{i=1}^{n}(X_i - \mu_X)^2}\sqrt{\sum_{i=1}^{n}(Y_i - \mu_Y)^2}} \tag{13}$$

4.3 RPC Two-Dimensional Projection of Data

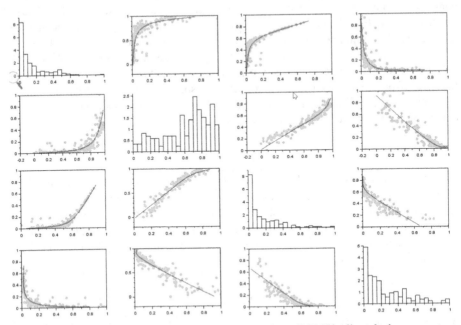

Fig. 1. RPC two-dimensional projection of performance data of 12345 hotline telephone operators in one month of 2019 in one district of Beijing

Figure 1 is the RPC two-dimensional projection of performance data of 12345 hotline telephone operators in one month of 2019 in one district of Beijing. Green points are observation data and red curves are RPC two-dimensional projection. According to all the score data, RPC learns skeleton structure of data distribution, so the ranking of performance of 10 telephone operators are obtained and the ranking of all the telephone operators in the evaluation system is decided. In consideration of multifarious operations, all the cases are summed and are collectively called the sum of handled cases, and the municipal checking number, the number of cases filed due to damage to public facilities, the number of garbage sorting sites are summed and collectively called the workload. All the indicators summed are calculated using Pearson correlation coefficient, and then four indicators which make the greatest contribution to ranking are chosen, including the total amounts of the cases handled, telephone traffic, workload and working hours. Therefore, the choice of indicators can achieve the aim of reducing the indicator dimension and computation complexity.

4.4 Ranking Result and Analysis

In this part, with the performance evaluation data of telephone operators, their performance evaluation ranking is conducted using ranking principal curve model, Elmap model and traditional weighting method. From Fig. 1, there are linear relation and non-linear relation on the ranking principal curves learned by the observation data based on the four indicators including the total amounts of the cases handled, telephone traffic, workload and working hours. The monotonous identity of this issue is $\alpha = [1, 1, 1, 1]^T$ which is learned from observation data and indicates the monotonous direction between observation data and ranking scores. Two -dimensional projection shows functional relationship of each evaluation indicator and relationship between evaluation indicators and ranking scores. If one telephone operator has high observation value in the four indicators of filed cases, telephone traffic, workload and attendance, his performance ranks higher in this month.

Table 1. The comparison of performance ranking of 12345 hotline telephone operators in January, 2019 of a district in Beijing

Name	Sum of handled cases	Telephone traffic	Workload	Working hours	Weighting ranking		Elmap ranking		RPC ranking	
					Score	Ranking	Score	Ranking	Score	Ranking
Miss Yang	6998	3009	317910	119	0.7459	1	0.8742	1	0.9937	1
Miss Pang	6380	6870	368090	144	0.7207	2	0.8531	2	0.9441	2
Miss Wang	6474	4932	328330	143	0.6473	3	0.7983	4	0.7174	3
Miss Xing	6459	5757	303550	142	0.6029	4	0.8327	3	0.7084	4
Miss Cao	5657	3492	291230	145	0.5646	5	0.7321	6	0.2337	8
Miss Bai	5984	4116	282350	142	0.5603	6	0.7645	5	0.3897	6
Miss Wu	5487	5158	284800	144	0.5593	7	0.6421	7	0.3421	7
Miss Zhai	6500	6279	270510	140	0.5472	8	0.3645	9	0.6811	5
Miss Liang	5870	3236	281930	149	0.5316	9	0.1375	10	0.1746	9
Miss Li	5316	2985	236390	141	0.4699	10	0.4565	8	0	10

From Table 1 it can be seen that the result using principal curve for ranking is close to the result using weighting for ranking and the result using Elmap model for ranking. The traditional evaluation adopts the method of weighted average. In different evaluation methods the evaluated indicators are different. As for the same evaluated indicators, the weights at different levels are determined by the administrative staff according to their experience, and different weights can generated different ranking results. Therefore, the traditional ranking is subjective. Elmap model has no explicit ranking rules expression, and the complexity of model is hard to determine although it can describe nonlinear data skeleton. In Table 1, Miss Cao's sum of handled cases and telephone traffic are both less than Zhai's and her telephone traffic is much less than Zhai's, but her weight is ranked fifth just because her workload and working hours are a little more than Zhai's.

However, in terms of data, Zhai's two pieces are much higher than Cao's, and in popular conception her weight should rank higher than Cao's, but her weight is only ranked the eighth. RPC method describes distributed skeleton structure of performance data through principal curve to rank telephone operators, which does not set weight value subjectively, and is in line with popular perception, so the principle of "letting data do the work" is implemented. This method is more scientific and interpretable than the method of weighted average and Elmap model.

5   Conclusion

The paper first normalizes the data, preprocesses the data using interpolation calculation, and chooses several ranking indicators with the highest relevancy degree to evaluation result using Pearson correlation coefficient. Then the RPC two-dimensional projection is conducted on evaluation indicators, and the ranking result of multiple indicator observation data is obtained with RPC model. The experiment is done with the performance data of 12345 hotline telephone operators in one month of 2019 in one district of Beijing, and the ranking result is in line with the popular perception. The experimental result proves that as for the ranking in performance evaluation, RPC method is more scientific and interpretable than the method of weighted average and Elmap model. In the future work, scalar indicators of unsupervised ranking learning result should be sought and the data with complex structure should be searched for to examine the effectiveness of RPC method.

Acknowledgements. This work is supported by National Key R&D Program of China (Grant number 2018YFC0704800), Natural Science Foundation of China (No: 41971366, 91846301), the 2018–2019 Excellent Talents Program in Xicheng District of Beijing and Beijing Urban Governance Research Center.

References

1. Hu, B.-G., Wang, Y., Yang, S.H., Qu, H.B.: How to increase the transparency of artificial neural networks? Pattern Recogn. Artif. Intell. **20**(1), 72–84 (2007)
2. Hu, B.-G., Qu, H.B., Wang, Y., Yang, S.H.: A generalized-constraint neural network model: associating partially known relationships for nonlinear regression. Inf. Sci. **179**, 1929–1943 (2009)
3. Kim, J.-O.: Multivariate analysis of ordinal variables. Am. J. Sociol. **81**(2), 261–298 (1975)
4. Potharst, R., Bioch, J.C.: Decision trees for ordinal classification. Intell. Data Anal. **4**(2), 97–111 (2000)
5. Klementiev, A., Roth, D., Small, K., Titov, I.: Unsupervised rank aggregation with domain-specific expertise. In: Proceedings of IJCAI 2009, pp. 1101–1106 (2009)
6. Unnikrishnan, R., Pantofaru, C., Hebert, M.: Toward objective evaluation of image segmentation algorithms. IEEE Trans. Pattern Anal. Mach. Intell. **29**(6), 929–944 (2007)
7. Rand, W.M.: Objective criteria for the evaluation of clustering methods. J. Am. Stat. Assoc. **66**(336), 846–850 (1971)
8. Freund, Y., Iyer, R.D., Schapire, R.E., Singer, Y.: An efficient boosting algorithm for combining preferences. J. Mach. Learn. Res. **4**, 933–969 (2003)

9. Webb, A.R.: Statistical Pattern Recognition, 2nd edn. Wiley, Chichester (2002)
10. Volkovs, M.N., Zemel, R.S.: New learning methods for supervised and unsupervised preference aggregation. J. Mach. Learn. Res. **15**, 1135–1176 (2014)
11. Lin, S.: Rank aggregation methods. Wiley Interdisc. Rev. Comput. Stat. **2**(5), 555–570 (2010)
12. Zhou, D., Weston, J., Gretton, A., Bousquet, O., Schölkopf, B.: Ranking on data manifolds. In: Thrun, S., Saul, L., Schölkopf, B. (eds.) Advances in Neural Information Processing Systems, vol. 16. MIT Press (2004)
13. Xu, B., et al.: Efficient manifold ranking for image retrieval. In: Proceedings of the 34th Annal Internatioanl ACM SIGIR Conference on Research and Development in Information Retrieval, pp. 525–534. ACM (2011)
14. Dekel, O., Manning, C.D., Singer, Y.: Log-linear models for label ranking. In: Advances in Neural Information Processing Systems, vol. 16 (2003)
15. Lin, F., Cohen, W.W.: The multirank bootstrap algorithm: semisupervised political blog classification and ranking using semi-supervised link classification. In: ICWSM (2008)
16. Harvelin, K., Kekalainen, J.: IR evaluation methods for retrieving highly relevant documents. In: Proceedings of the 23rd Annal Internatioanl ACMSIGIR Conference on Research and Development in Information Retrieval, pp. 41–48. ACM (2000)
17. Baeza-Yates, R., Ribeiro-Beto, B.: Information Retrieval. Addison Wesley, Boston (1999)
18. Chapelle, O., Metlzer, D., Zhang, Y., Grinspan, P.: Expected reciprocal rank for graded relevance. In: Proceedings of the 18th ACM Conference on Information and Knowledge Management, pp. 621–630. ACM (2009)
19. Volkovs, M.N., Zemel, R.S.: BoltzRank: learning to maximize expected ranking gain. In: Proceedings of the 26th Annual International Conference on Machine Learning, pp. 1089–1096. ACM (2009)
20. Volkovs, M.N., Zemel, R.S.: A flexible generative model for preference aggregation. In: Proceedings of the 21st International Conference on World Wide Web, pp. 479–488. ACM (2012)
21. Schölkopf, B., Simard, P., Smola, A.J., Vapnik, V.: Prior knowledge in support vector kernels. In: NIPS, pp. 640–646 (1998)
22. Vapnik, V., Vashist, A.: A new learning paradigm: learning using privileged information. Neural Netw. **22**(5), 544–557 (2009)
23. Li, C.-G., Mei, X., Hu, B.-G.: Unsupervised ranking of multi-attribute objects based on principal curves. IEEE Trans. Knowl. Data Eng. **27**, 3404–3416 (2015)
24. Chen, Q., Sun, Y.: Analysis of the significance of performance appraisal in enterprise salary management. Mod. Market. **12**, 195–196 (2019)
25. He, W.Q.: Discussion on the problems and solutions of employee performance appraisal. Times Financ. **03**, 62–63 (2020)
26. Du, J.: The practice of Technology Governance in the transformation of service oriented government. J. Southwest Univ. **45**(06), 37–44+201–202 (2019)

Proximity-Based Aggregation Method for LBS Human Mobility Data

Yao Long[1], Lin Zhang[2], and Tao Yuan[1(✉)]

[1] China University of Geosciences (Beijing), Beijing 100083, China
ly072511@163.com, yuantao@cugb.edu.cn
[2] Hunan Key Laboratory of Land Resources Evaluation and Utilization, Changsha 410118, Hunan, China

Abstract. Human mobility is an inevitable element of urban development. The rise of big data has accelerated the popularity of various location-based service (LBS) data, and mobility data has gradually become the mainstream data. When mobility data is applied to the research into human mobility, geographic location will be integrated into a complex network structure composed of massive human mobility data to form a geographic network space, which will integrate human mobility and urban development in the era of big data. Mobility data features large quantity, complexity and redundancy, and its aggregation is valuable. In this paper, based on LBS human mobility data, a geographic proximity-based aggregation method for mobility data is proposed, aiming at aggregating mobility data of geographic proximity according to human mobility values to achieve mobility data aggregation in geographic sense without changing the situation of data, and generate a series of urban aggregation areas formed by closely approximate internal mobility data.

Keywords: Geographic location · Human mobility · Mobility data · Aggregation

1 Introduction

In the process of urban development, exchanges of population and material resources take place between cities. Human mobility refers to various short-term, repetitive or periodic movements of population between regions, which is closely related to time. Human mobility in relatively short cycle is one of the main driving forces for urban sprawl [1–3], and it represents the interconnection between cities and reflects the real-time exchanges of human mobility information between cities as well as the influence of a city on surrounding cities. The distribution pattern of human mobility determines the characteristics of urban spatial structure evolution to a large extent, and directly affects urban spatial efficiency and sustainable development [4].

In contrast, due to system differences, in western countries, there is no such concept as "migrant population" [5], but groups such as international migrants and ethnic minorities in the west have similar socio-economic characteristics with "migrant population" in China [6]. Western scholars focus more on spatial isolation and integration, and they

X. Meng et al. (Eds.): SpatialDI 2020, LNCS 12567, pp. 218–230, 2021.
https://doi.org/10.1007/978-3-030-69873-7_16

study the spatial distribution of population in cities from the aspect of the distribution of international migrants and ethnic minorities in big cities and megacities [7–9]. Since the early 1990s, some Chinese scholars have also begun to focus on the distribution and influencing factors of human mobility in big cities and megacities [10–14], and have conducted a lot of researches on the characteristics [15], influencing factors and migration mechanism [16] of urban human mobility in China. The research on the spatial distribution of human mobility mainly is mainly the macro analysis of provincial administrative regions [17, 18] with net immigration rate [19, 20], migration scale [21] and other single indicators as the basis for classification. As a result, such research is rather subjective and not so comprehensive and persuasive. With districts and counties as the basic units for spatial analysis, Liu [22] uses census data to build a composite index system for regional division of migrant population, and has found the law governing regional differences of the spatial distribution of migrant population in China. In general, relevant research has found that migrant population in China is mainly concentrated in the Yangtze River Delta, Pearl River Delta, Beijing, Tianjin, Hebei and other coastal urban agglomerations, especially megacities there [17, 21, 23–25].

From the perspective of urban geography, urban spatial organization is gradually shifting to a multi-center mode with the change of geographical distance, and urban development tends to feature a grid-based layout. On the basis of the "flow space" of urban development, this paper proposes the concept of "urban network" which pays more attention to spatial connection. Urban geography begins to change from paying attention to the urban hierarchical scale system to the research into urban network structure, and the organization mode has changed from the central place mode to the multi-center and mobile network mode. As a key component of urban network structure, mobility data has become one of the main types of data related to urban development [1–3] with the maturity of data mining technology, the continuous development of data and the increase in data types. Human mobility data, as a specific type of geographic Origin-Destination mobility data, is not only a special trajectory data, but also a special spatial interaction data (SI). Each object of mobility data consists of a series of time and a pair of position points. The geographical positions of the starting point and the ending point are usually points. Some data can be converted into regions through clustering, but the trajectory is arbitrary. The visualized map of mobility data is the most common way to present mobility data because the stream is represented by straight lines or curves connecting the start and end points. The organization structure of location-based mobility data is commonly used to understand the location characteristics and spatial patterns of steaming data [26, 27]. The visualized map of mobility data shows the links between locations and display patterns of geographic movement [28]. For a long time, flow mapping has been widely used to map human migration [27, 29], commodity traffic [30], transportation [31], commuting patterns [32] and crime patterns [33].

The existing visualized map of mobility data is only effective in depicting small data sets. As for mobility data with millions of pairs of starting and ending points, the visualized map will become difficult to recognize quickly with the increase in data size, and the serious problem of confusion will be caused due to a large number of intersections and trajectory overlaps. At present, many new mobility data mapping methods have been proposed to visualize and discover large-scale mobility data integration patterns [26, 34–37]. To reduce the confusion of visualized map of mobility data, we can minimize the crossing parts of mobility data through location aggregation points and their neighboring points [34], or through edge re-planning [26] and edge bundling (EB) [36–38], which is essentially similar to other drawing methods. A variety of methods can be used to realize the interactive visualization of spatial mobility data with non-spatial views, such as ordered matrix [39], maps, and innovative combinations of matrices and other methods including Map2 [40], interactive OD map [41] and exploratory visualization [42].

"Tencent Location Big Data" is a big data platform for daily human mobility based on location-based service (LBS) launched by Tencent. With the popularization of intelligent communication devices and the continuous development of spatial location technology, it has been widely used in various fields [43]. With the platform, users' geographic location, preferences, travel routes, activity traces, social patterns and other information can be found out to reveal users' daily spatio-temporal behavior. Internet big data has become an important means to reflect public social activities [44]. These geo-behavioral big data are collected from various sources. They are real-time, objective and easy to be analyzed and predicted, which makes up for the defects of conventional means of survey such as questionnaire, sampling and census, and provides sufficient and highly spatially and temporally accurate measured mobility data for the online research into human mobility. The dimension of geographical location has established a bridge between the real world and online social networks to share users' current location and activity information [45].

This paper uses LBS's human mobility data collected in 365 cities across the country based on Tencent Location Big Data to replace the geographical location of the starting and ending points of mobility data with regional units, and effectively aggregates mobility data iteratively in geographical sense based on geographical location proximity. By selecting target cities and setting different iteration ratios, mobility data are spatially aggregated to improve the complex and disordered distribution status of mobility data and better visualize the spatial structure of urban network based on mobility data. The aggregation effect of different target cities is compared and analyzed to explore urban network structure.

2 Methods

2.1 Data Preparation

The method proposed in this paper is aggregation based on human mobility data, which is a series of records with origin and destination points. The weight value of mobility data is the human mobility value between the two points. The origin and destination point of each record are marked with specific geographical locations. The mobility data used in this paper is the human mobility data based on Tencent Location Big Data, and its origin and destination point is determined according to its geographic location in a certain prefecture-level city.

The specific geographic location determines the neighboring nodes of each node through geographic proximity. This paper uses the principle of whether two regions have a common border to determine geographical proximity, namely, geographical proximity of two different prefecture-level cities can be determined by their common border.

Therefore, the original input data of the algorithm is the proximity between the original human mobility data (OD) and the geographic location.

2.2 Principle

Set Iterative Ratio r. List the human mobility values of OD in descending order. As is shown in Fig. 1, the data is generally distributed in a heavy-tailed manner with the mobility data of large human flow values accounting for a very small proportion since the starting and ending points of human mobility data are fixed geographical locations and are represented by specific prefecture-level cities. According to the first 1,000 pieces of human mobility data, it is found that for every 100 pieces increased, the increase in the number of cities will slow down with the increase in mobility data (see Fig. 2). According to "2019 City Business Glamour Ranking" released by China's New First-tier City Research Institute, cities at or above the fourth level (209 administrative units at the prefecture level) account for 62.01% of all the 337 cities at or above the prefecture level in the ranking. As can be seen from Fig. 2, when the first 500 pieces of data are extracted, the number of cities related to the mobility data is 218, close to that of cities at or above the fourth level in the ranking, and most major cities across the country have been covered.

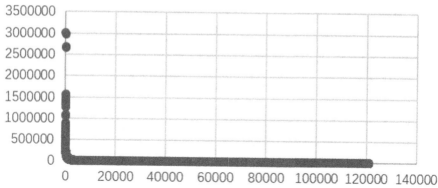

Fig. 1. Distribution of human mobility data. The x-axis of the scatter plot is the number of mobility data, and the y-axis is the value of the mobility data.

The relationship between two city nodes is the basis for judging mobility data aggregation, which will inevitably bring about changes in city nodes. Therefore, the aggregated city nodes should be selected first. Once a certain proportion of mobility data is selected as the basis of iteration, the change of proportion will represent the change of iterative target area. Set the variable: iteration proportion (r). Because there are different levels of cities, in order to find more accurate aggregation methods for more effective aggregation effects, set the iteration proportion of $r = 100$, $r = 200$, $r = 300$, $r = 400$ and $r = 500$ as the basis for the iterative aggregation of mobility data.

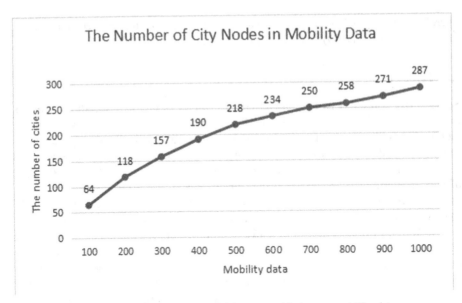

Fig. 2. Line chart of number of cities covered in human mobility data.

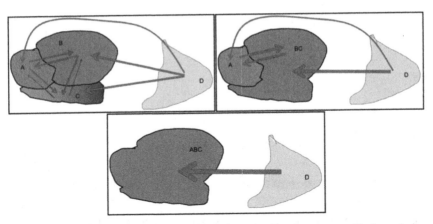

Fig. 3. Schematic diagram of mobility data aggregation based on geographical proximity.

Traverse OD Line by Line. Judging by proximity information, determine the geographical position relation of the dotted points in each record. If the starting and ending points are geographically adjacent, the two points are viewed as in one area. Update the mobility data of people migrating from a certain point in the research area to the two points. The combined population flow value is the human mobility data from that point to the research area. Otherwise, update the data with the two points as the starting points (see Fig. 3). After updating the mobility data, continue to update the proximity information of geographic location which serves as the basis for processing the next record.

Extract Result. Extract the last updated OD and the merged information. When judging whether two points are in proximity relationship, attention should be paid to judging whether the points have been merged as one region, and then use the updated OD and the merged information in a timely manner.

3   Results

As is shown in Fig. 4, after visualizing the data with a human mobility value of greater than 100 person-times, we find that the visualized mobility data roughly show that China's human mobility is mainly concentrated in the southeast wall of the Hu's Line, and that its spatial distribution is disordered and mixed. In order to better display the spatial structure of urban geographic network, it is necessary to aggregate the mobility data.

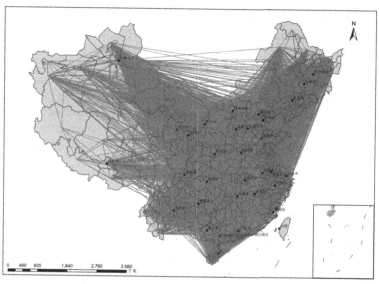

Fig. 4. Visualization of initial human mobility data (with human mobility value greater than 100 person-times).

3.1   The Aggregation Effect of Mobility Data is Obvious, and the Urban Network Structure is Obvious

Apply different iteration ratios (r) to mobility data aggregation methods based on geographic proximity, the results shown in Fig. 5 is obtained and visualized by using natural discontinuity classification display method used in Fig. 4. The results show the main streaming patterns and important locations/regions in the data. Less mobility data is used to show the spatial structure of the data without losing data. Merge city nodes into regions based on their geographical locations, and replace human mobility between cities with that between regions to directly show the current situation and influence scope of human mobility between regions. Human mobility between cities follows the distribution patter of Hu's Line with dense population in the east and sparse population in the west, and higher human mobility value in the east and lower in the west.

With the increase of the iteration ratio "r", the mobility data of merged between target cities gradually increase, the merging effect of mobility data becomes remarkable, and the urban backbone network is basically formed. The network of human mobility cities in our country is mainly a rhombus structure with Beijing-Tianjin-Hebei region, Yangtze River Delta region, Pearl River Delta region and Sichuan-Chongqing region as the four vertices, which is closely related to the advantageous geographic location and economic development in southeast China. Secondly, when cities are aggregated into regions, they are more directly related to human mobility in neighboring cities, which directly reflects the radiation and driving effect of a region (or city group) on the neighboring areas and shows an obvious core-edge structure. The aggregation effect of mobility data shows the rhombus spatial structure in China's urban network. Especially when the ratio "r" is 200, the spatial pattern becomes more obvious, and the four vertex regions have the strongest attraction to human mobility. As the intersection of the diagonal lines of the rhombus structure, Wuhan enjoys superior geographical location and more convenient traffic condition. Compared with the visualization effect when "r = 100", the result of "r = 200" shows that Wuhan attracts and gathers its neighboring areas to a certain extent. The four vertex regions, as the main growth poles, have larger radiation range with the increase of the iteration ratio. Among all the regions, the Yangtze River Delta region has the fastest growth in radiation range. The comparison of the results of "r = 300" and "r = 400" shows that the Yangtze River Delta region has absolute attraction to cities nationwide, but the degree of attraction decreases with the increase of geographical distance.

When the r equals 100, Urumqi forms an urban network with other city nodes in its original state of city nodes. However, when r equals 200, Urumqi and Hui Autonomous Prefecture of Changji are merged into one region, which forms an urban network with other city nodes. With the increase of the number of designed cities, they have more human mobility with the neighboring cities and more intensive exchanges with the coastal areas in the southeast, which directly shows Urumqi's dominant position in the northern slope of Tianshan Mountains. Besides, Xining and Lanzhou regions have gradually become important hubs in northwest inland and the coastal areas in the southeast with their strengthened gathering force. Therefore, effective mobility data aggregation has highlighted the population aggregation role played by northwest inland cities in the human mobility network.

There is a high degree of cohesion in the southwest region. The visualization effect of the directional mobility data of central Yunnan and Sichuan and Chongqing regions is obvious, and Guangxi region seems to have become a part of the spatial structure of the urban network as the fifth growth pole. When "r" equals 500, such aggregation effect becomes the most prominent, indicating that the human mobility in Guangxi province occupies a relatively large proportion in the human mobility network. The southern Fujian region is closely linked with the Yangtze River Delta region and the Pearl River Delta region, and thus the long-distance human mobility occupies a relatively small

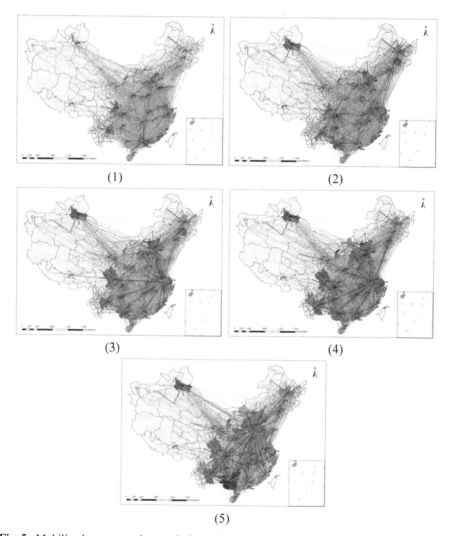

Fig. 5. Mobility data aggregation results by applying different iteration ratio of "r" (the respective iteration ratio of (1) to (5): r = 100, r = 200, r = 300, r = 400, r = 500).

proportion. In the three northeastern provinces, a belt-shaped region centered on Harbin, Changchun and Shenyang has been formed. These regions are closely connected with each other, and their human mobility with the Beijing-Tianjin-Hebei region is strong.

In a word, the visualization results of the aggregated mobility data are no longer overlapping and intersecting disordered mobility data, but highlight the spatial structure of China's urban network, as well as the accessibility of the network. Effective aggregation of mobility data can reflect the gradual development of China's urban network to a multi-center model, and the more obvious connection between regions.

3.2 The Larger Aggregation Area and More Obvious Urban Agglomeration Structure

With the increase of the iteration ratio "r", the number of target cities for mobility data aggregation increases, the regional area of aggregation results becomes larger (see Fig. 6), and even new aggregation regions appear, which reflects the evolution of urban agglomeration. The number of aggregated regions has increased from 17 to 28. These regions are mainly in part of the southeast, and are likely to cover all southeast regions. The main aggregated regions in the northwest region are Urumqi in the northern slope of Tianshan Mountain urban agglomeration, Xining and Lanzhou in Lan-Xi urban agglomeration, Yinchuan in Ningxia's urban agglomeration along the Yellow River, and Baotou in Hu-Bao-E-Yu urban agglomeration. With the increase of the iteration proportion, Lanzhou region keeps expanding to the southeast partly because of the construction of high-speed rail and the improvement of traffic conditions there. But Lanzhou region does not overlap Xining region because Xining is located on the edge of Qinghai-Tibet Plateau while Lanzhou is located on the Loess Plateau, and the different terrain conditions in the two plateaus may directly affect human mobility between the two cities and allow Lan-Xi urban agglomeration to develop in a "dual-core" model. Hu-Bao-E-Yu urban agglomeration mainly consists of four prefecture-level cities. Judging by the aggregation effect, the human mobility between Hohhot and Baotou is strong, and at the same time Ulanqab has been attracted to become a new aggregation unit.

With the increase of the iteration ratio "r", the area of Beijing-Tianjin-Hebei region, the Yangtze River Delta and Pearl River Delta regions has become increasingly large. When r = 300, the complete regional structure has been formed, which is related to the highly developed internal network of the urban agglomeration. When r = 500, the Beijing-Tianjin-Hebei region is linked to the central plains urban agglomeration mainly due to the rapid development of the traffic arteries in those regions. As a large transportation hub, Zhengzhou has become the main junction of the north-south and east-west trunk line, directly driving the main human mobility across the country and making the contact with the Beijing-Tianjin-Hebei region closer. With the increase of the iteration ratio, Sichuan-Chongqing urban agglomeration has incorporated more cities. Not only is Aba Tibetan and Qiang Autonomous Prefecture in Tibet but also the urban agglomeration in central Guizhou is incorporated, which demonstrates the sound development of Sichuan-Chongqing urban agglomeration in the southwest as the fourth growth pole of China's urban network as well as its strong attraction to and driving effect on the neighboring regions and its increasing radiation scope.

New elements keep being incorporated into the Wuhan and Changsha-Zhuzhou-Xiangtan regions which are connected with the Nanchang region to form an urban agglomeration in the middle reaches of the Yangtze River. The three aggregated regions in Shandong Peninsula region have become a whole region of Shandong Province with Jinan and Qingdao becoming the main driving forces for the development of their neighboring cities.

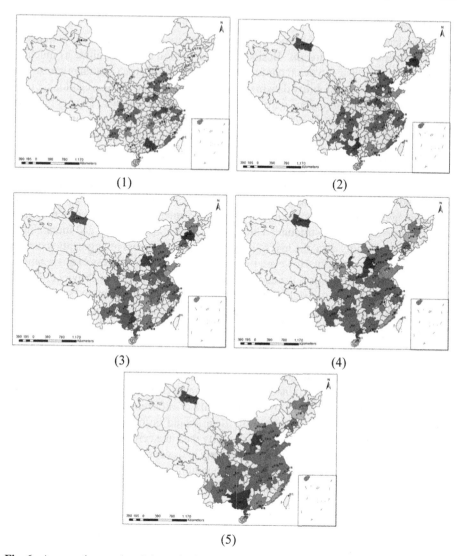

(1)

(2)

(3)

(4)

(5)

Fig. 6. Aggregation results of city nodes by applying different iteration ratio of "r" (the respective iteration ratio of (1) to (5): r = 100, r = 200, r = 300, r = 400, r = 500).

4   Conclusion

This paper proposes a mobility data aggregation method based on geographic proximity to solve the problem of the aggregation of LBS human mobility data. An iteration ratio "r" is set in the method based on the amount of the human mobility value to aggregate the mobility data to different degrees, which can not only effectively aggregate mobility data and hide the internal mobility data in the aggregated region, but also identify regions with different aggregation degrees in the city according to the location information of the aggregated region. After the data streams are merged, the amount of global visualized mobility data can be reduced without changing the distribution status of the data, and a visually clear visualized mobility data map will be created to summarize patterns in a large number of spatial streams. Besides, data resolution can be retained as much as possible in the map space to avoid confusion and reveal the abstract spatial organization and structure patterns of mobility data in the network. Moreover, the aggregation regions are multiple neighboring regions with close contact with one another. The aggregation in the regions is initially realized according to the number of the streams, and the result is approximately the same with the "5 + 9 + 6" urban agglomeration system designated by the state.

From the perspective of global mobility data, this method focuses on the geographic relationship between the starting and ending points of mobility data, and aggregates the starting and ending points according to the proximity relationship of geographic locations. The aggregation results are highly aggregated internally without compromising the original organizational structure. Through repeated loop iterations and hierarchical aggregation of mobility data, the loss of information in the process of aggregation can be reduced as much as possible. However, to use this method in the process of aggregation, accurate neighbor information is needed. But Zhoushan in Zhejiang Province, for an instance, does not have any neighboring prefecture-level city, so it cannot be iteratively aggregated into other regions. For such cities, mobility data aggregation should be conducted based on other neighbor relationships. Although the original human mobility data and the number of city nodes will decrease due to the aggregation of nodes in the iteration process, the iteration results will not be affected, and only the original human mobility data and the iteration ratio will affect the iteration results. The human mobility data used in this paper is determined based on the scale range of prefecture-level cities. The aggregation effect generated by mobility data of different scales varies from one to another. As a result, the target mobility data should be obtained by applying this method on demand. In this study, human mobility data at a certain time is chosen as the data source, but such data cannot comprehensively and accurately summarize the spatial structure and characteristics of the national urban human mobility network. Therefore, the data during a longer period of time should be chosen to improve the effectiveness of analysis.

Acknowledgements. This work was supported by the Open Topic of Hunan Key Laboratory of Land Resources Evaluation and Utilization, grant number SYS-MT-201901. We greatly appreciate the constructive comments and suggestions from reviewers.

References

1. Deng, Y., Si, Y.: The spatial pattern and influence factors of urban expansion: a case study of Beijing. Geogr. Res. **34**(12), 2247–2256 (2015)
2. Han, H., Yang, C., Song, J.: The spatial-temporal characteristic of land use change in Beijing and its driving mechanism. Econ. Geogr. **35**(05), 148–154+197 (2015)
3. Yuan, Y., Xu, X., Xue, D.: Spatial distribution, evolution and driving force of non-registered population of Guangzhou metropolitan area in 1990–2000. Econ. Geogr. (02), 250–255 (2007)
4. Li, J., Ning, Y.: Population spatial change and urban spatial restructuring in Shanghai since the 1990s. Urban Plann. Forum (02). 20–24 (2007)

5. Zhao, M., Wang, D.: The spatial sprawl and driving mechanism of the floating population in Beijing metropolitan areas. Scientia Geographica Sinica **39**(11), 1729–1738 (2019)
6. Zhao, M., Liu, S., Qi, W.: Comparison on foreign and domestic migrant communities: characteristics and effects. City Plann. Rev. **39**(12), 19–27 (2015)
7. Alba, R.D., Logan, J.R., Stults, B.J., et al.: Immigrant groups in the suburbs: a reexamination of suburbanization and spatial assimilation. Am. Sociol. Rev. **64**(3), 446–460 (1999)
8. Allen, J.P., Turner, E.: Ethnic residential concentrations in United States metropolitan areas. Geogr. Rev. **95**(2), 267–285 (2005)
9. Logan, J.R., Zhang, W., Alba, R.D.: Immigrant enclaves and ethnic communities in New York and Los Angeles. Am. Sociol. Rev. **67**(2), 299–322 (2002)
10. Bao, S., Zhang, B.: The spatial distribution and growing trends of migrant population in Beijing. J. Capital Normal Univ. (Nat. Sci. Ed.) **33**(02), 74–78 (2012)
11. Feng, J., Zhou, Y.: The growth and distribution of population in Beijing metropolitan area (1982–2000). Acta Geogr. Sin. **06**, 903–916 (2003)
12. Geng, H., Shen, D.: Spatial distribution and mechanism of migrant population in shanghai. City Plann. Rev. **12**, 21–31 (2009)
13. Yi, C., Gao, B., Huang, Y.: Spatial distribution of Beijing's Hukou-residence separated population and its determinants. Chin. J. Popul. Sci. (01), 33–43+126–127 (2014)
14. Jie, Z., Xiao, L., Bo, X.: Spatial distribution and variation of floating population in megacities --case study of Beijing, Shanghai, Guangzhou and Wuhan. Urban Plann. Forum (06), 56–62 (2015)
15. Chen, T., Liu, S.: Research summary of the floating population in China. J. Anhui Agric. Sci. **37**(30), 14940–14942+14948 (2009)
16. Gu, C., Cai, J., Zhang, W., et al.: A study on the patterns of migration in Chinese large and medium cities. Acta Geographica Sinica (03), 14–22 (1999)
17. Ding, J., Liu, Z., Cheng, D., et al.: Areal differentiation of inter-provincial migration in China and characteristics of the flow field. Acta Geogr. Sin. **60**(01), 106–114 (2005)
18. Liu, Y.: The spatio-temporal distribution of floating population and its development tendency in China. China Popul. Resour. Environ. **01**, 139–144 (2008)
19. Lu, Q., Wang, G., Yang, C., et al.: Some explanations to the relationships between the geographical distribution change of migrants and economic development change in regions of China, 1990 and 2000. Geogr. Res. (05), 765–774+949 (2006)
20. Lu, Q., Wu, P., Lu, L., et al.: The relation between the characteristics of the migrants and the economic development in Beijing and the regional differentiation of their distribution. Acta Geogr. Sin. (05), 851–862 (2005)
21. Yu, T.: Spatial-temporal features and influential factors of the China urban floating population growth. Chin. J. Popul. Sci. (04), 47–58+111–112 (2012)
22. Liu, S., Deng, Y., Hu, Z.: Research on classification methods and spatial patterns of the regional types of China's floating population. Acta Geogr. Sin. **65**(10), 1187–1197 (2010)
23. Shen, J.: Changing patterns and determinants of interprovincial migration in China 1985–2000. Popul. Space Place **18**(3), 384–402 (2012)
24. Wang, Y., Li, H., Yu, Z., et al.: Approaches to census mapping: Chinese solution in 2010 rounded census. Chin. Geogra. Sci. **22**(3), 356–366 (2012)
25. Cao, G., Liu, T.: Rising role of inland regions in China's urbanization in the 21st century: the new trend and its explanation. Acta Geogr. Sin. **66**(12), 1631–1643 (2011)
26. Phan, D., Xiao, L., Yeh, R., et al.: Flow map layout. In: IEEE Computer Society, Los Alamitos, pp. 219–224 (2005)
27. Tobler, W.R.: Experiments in migration mapping by computer. Am. Cartogr. **14**(2), 155–163 (1987)
28. Guo, D., Zhu, X., Jin, H., et al.: Discovering spatial patterns in origin-destination mobility data. Trans. GIS **16**(3), 411–429 (2012)

29. Tobler, W.R.: A model of geographical movement. Geogr. Anal. **13**(1), 1–20 (1981)
30. Ullman, E.L., Dacey, M.F.: The minimum requirements approach to the urban economic base. Pap. Reg. Sci. **6**(1), 175–194 (1960)
31. Gould, P.R.: On the geographical interpretation of eigenvalues. Trans. Inst. Br. Geogr. **42**, 53–86 (1967)
32. Corcoran, J., Chhetri, P., Stimson, R.: Using circular statistics to explore the geography of the journey to work*. Pap. Reg. Sci. **88**(1), 119–132 (2009)
33. Groff, E., McEwen, T.: Exploring the spatial configuration of places related to homicide events. Institute for Law and Justice, Alexandria, VA (2006)
34. Andrienko, N., Andrienko, G.: Spatial generalization and aggregation of massive movement data. IEEE Trans. Visual Comput. Graphics **17**(2), 205–219 (2011)
35. Guo, D.: Flow mapping and multivariate visualization of large spatial interaction data. IEEE Trans. Visual Comput. Graphics **15**(6), 1041–1048 (2009)
36. Holten, D., Van Wijk, J.J.: Force-directed edge bundling for graph visualization. Comput. Graph. Forumz **28**(3), 983–990 (2009)
37. Verbeek, K., Buchin, K., Speckmann, B.: Flow map layout via spiral trees. IEEE Trans. Visual Comput. Graphics IEEE Trans. Visual Comput. Graphics **17**(12), 2536–2544 (2011)
38. Cui, W., Zhou, H., Qu, H., et al.: Geometry-based edge clustering for graph visualization. IEEE Trans. Visual Comput. Graphics **14**(6), 1277–1284 (2008)
39. Guo, D.: Visual analytics of spatial interaction patterns for pandemic decision support. Int. J. Geogr. Inf. Sci. **21**(8), 859–877 (2007)
40. Guo, D.S., Chen, J., MacEachren, A.M., et al.: A visualization system for space-time and multivariate patterns (VIS-STAMP). IEEE Trans. Visual Comput. Graphics **12**(6), 1461–1474 (2006)
41. Wood, J., Dykes, J., Slingsby, A.: Visualisation of origins, destinations and flows with OD maps. Cartogr. J. **47**(2), 117–129 (2010)
42. Yan, J., Thill, J.C.: Visual data mining in spatial interaction analysis with self-organizing maps. Environ. Plan. B-Plan. Des. **36**(3), 466–486 (2009)
43. Bie, W., Yi, Z., Xiao, W., et al.: Urban community structure mining based on Weibo Check-in data. Geomatics World **26**(04), 68–73 (2019)
44. Qin, X., Zhen, F., Xiong, L.: Methods in urban temporal and spatial behavior research in the Big Data Era. Prog. Geogr. **32**(09), 1352–1361 (2013)
45. Dan, J.: Thoughts and several suggestions on LBS development in China. Geomatics World **10**(06), 22–24 (2012)

Research on the Division of Functional Zones in Downtown Beijing Under the Background of Big Data

Te Qi, WeiNa Zhang, and Tao Yuan(⊠)

China University of Geosciences (Beijing), Beijing, China
yuantaobj@qq.com

Abstract. An urban functional area is a regional space that realizes the spatial aggregation of relevant social resources and fulfills a specific urban function. As a form of modern urban development, it is an important spatial carrier to fulfill the economic and social functions of a city, and centrally reflects the characteristics of a city. The identification and subsequent analysis of urban functional areas will help people find out the problems in the development of urban space, and are of great significance to urban planning, regional economic development, optimization of urban functional areas and the promotion of urbanization. In this paper, the functional areas of downtown Beijing are divided and the results of division are analyzed based on the hourly statistics of mobile phone signaling data collected on April 17, 2019 (Sunday) and April 20, 2019 (Wednesday), the POI data of Beijing in 2018, and the traditional data of the boundary of Beijing's central urban area.

In this paper, the supervised classification and unsupervised classification methods are adopted to identify and compare the functional areas in downtown Beijing. To apply the supervised classification approach, the random forest classification algorithm is used to analyze mobile phone signaling data. While the ISO clustering unsupervised classification approach is used to analyze the mobile phone signaling data. The functional area classification diagram of downtown Beijing is drawn through both supervised classification and unsupervised classification.

Finally, in this paper, the results of supervised classification and unsupervised classification are compared and analyzed to find out the advantages and disadvantages of the two approaches, and compares them with the remote sensing image map of Beijing. It is found that there is still room for further development within downtown Beijing and that there are structural problems in the functional areas of downtown Beijing.

Keywords: Mobile phone big data · Signaling data · POI data · Random forest · Supervised and unsupervised classification

1 Introduction

With the continuous urbanization, cities no longer just satisfy residents' daily needs and habitat needs. Instead, cities are now developing to fulfill diversified functions,

© Springer Nature Switzerland AG 2021
X. Meng et al. (Eds.): SpatialDI 2020, LNCS 12567, pp. 231–244, 2021.
https://doi.org/10.1007/978-3-030-69873-7_17

and thus the division of functional areas within a city has become more clear to meet residents' higher demands for work and life. So the concept of urban functional area is proposed. Real-time and accurate identification and division of urban functional areas are of great significance to the rational urban layout, the sound economic development and environmental protection of a cityQuery.

As big data is widely applied, scholars apply data of different scales and from various sources to identify and divide urban functional areas. For example, Shuang Ma and Long Ying et al. used more than 43 million desensitized car hailing records collected from Didi, the world's largest online car-hailing platform, to identify China's functional urban areas with fine-grained features nationwide. They call this a new method for defining urban functional areas in China. Chi Jiao and Jiao Limin et al. carried out quantitative identification and visualization of urban functional areas based on POI data, reclassified POI data, identified single functional areas and mixed functional areas of the city with the reclassified data, designated single functional areas by a certain color, and used RGB additive color method to visualize mixed urban functional areas. Yu Lu and He Xiang et al. conducted recognition research on urban functional areas based on spatio-temporal semantic mining. With rectangular areas of the city as the research sample and buildings as the basis for classification, they divided the research sample into effective basic areas. Then, they conducted spatio-temporal semantic mining based on the location check-in data on Weibo.com and POI (Points of Interest) data in each basic area, and adopted Dirichlet Multinomial Regression (DMR) as theme model to generate regional functional vector. Finally, through vector clustering, they conducted functional identification of the areas according to the proportion of POI categories.

At present, there is still room for improvement in the completed research. Firstly, POI data has some limitations in application. When it is used for classification, large, medium and small cities are different in terms of city size, transportation facilities and population distribution. Therefore, when POI data is used for identification and classification, such differences need to be considered. For cities of different sizes, more researches into the improvement of the accuracy of classification are needed. Second, only a single type of data is introduced in the research, so the data is relatively single and this may lead to structural differences. Third, each POI point has its corresponding function, but the corresponding weight of each function is not considered. For example, in areas such as subway stations and railway stations where transportation functions are greater than other functions, the corresponding weight of POI points must be much greater than that of the surrounding stores, and the identification result should be traffic land. However, since there are too many stores around these places, the unweighted railway station is just one of the many POI points, so the corresponding weight should be considered when people classify with POI points to improve classification accuracy.

In this paper, mobile phone signaling data and POI data are used and supervised classification and unsupervised classification approaches are adopted to identify and divide the functional areas in downtown Beijing. Then, in this paper, the advantages and disadvantages of the two approaches in data classification are compared. Finally, in this paper, the functional areas in downtown Beijing are analyzed and evaluated and suggestions for improvement are put forward.

2 Research Areas and Data Source

In this paper, the most developed and representative downtown Beijing is chosen as the research area, including Chaoyang District, Haidian District, Dongcheng District, Xicheng District, Fengtai District and Shijingshan District. Within these districts, there are 101 streets and 31 regions. The mobile phone signaling data was recorded every other hour at each location on April 17, 2019 (Wednesday) and April 20, 2019 (Saturday) provided by China Unicom, totaling 860,000 pieces. The POI data is Beijing's POI interest point data in 2018, which includes the name of each POI point, primary and secondary classification type, longitude and latitude coordinates, administrative jurisdiction, subdistrict, address, postal code and other geographic attributes. In addition to the above research data, in this paper, the remote sensing data of Beijing generated by Google Earth are compared and analyzed.

3 Research methods

3.1 Random Forest Classification Algorithm

Random forest is a machine learning model and a means of supervised classification. It name suggests that the algorithm builds a forest randomly, which is composed of several decision trees. First, the algorithm converts all the features in the data into float form, selects m samples from the sample set as training sample set, and the remaining samples serve as test sample set, so as to evaluate the accuracy of the algorithm. N represents the total number of features. Among all the features, this algorithm chooses n (n is fewer than N) features as the number of features in each decision tree, and each decision tree judges the data splitting mode according to its own n features. The final result is produced through majority voting by all decision trees.

Random forest method has many advantages in data classification and division. Compared with other algorithms, it has two advantages. Firstly, it has two random introduction: one is that there are put back samples during the selection of training data sets, and the other is that the characteristics of decision trees are randomly selected, which can greatly reduce the interference of noise points in data. Secondly, it adapts to data sets well and there is no need to manually select its characteristics. Thirdly, the mutual influence between each characteristics can be detected during algorithm training.

Density analysis refers to the calculation of data aggregation of the whole research area based on the input factor data by summing the point features that fall within each raster mobile. In order to improve the accuracy of random forest model, in this paper, the network density of POI point data is analyzed. Firstly, input POI points into the grid of the research area, create fishing nets with each POI point viewed as a small functional unit, and then count the number of POI points falling into each grid pixel. The more POI points that fall into the grid, the more centralized the urban function of the area. The results of network density analysis of POI points are proposed as a feature of random forest model to increase its accuracy.

3.2 ISO Clustering Analysis

ISO clustering analysis is a kind of unsupervised classification method which is in fact an iterative optimization clustering process. This method is also called mobility average value method. The prefix "iso" in "isodata" clustering algorithm is an abbreviation of the iterative self-organizing method that performs clustering operation. The clustering classification realized by this method is done in the following steps: for each iteration of the algorithm, all sample data will be distributed to the clustering center of the current classification to calculate the new average value of each class. Generally speaking, the researchers do not know in advance how many classes the unsupervised classification method will eventually produce. Therefore, they need to select the number of classes greater than the normal number according to common sense, and then reselect the number of classes according to the generated classification results and compare the accuracy of the results. ISO clustering analysis is generally an iterative process realized through algorithms, and the minimum Euclidean distance is calculated when assigning each sample data to its respective clustering center. When using this method, the researchers need to specify a clustering number, and then assign samples to the nearest clustering center for each following clustering operation. Then, according to the attribute average value of each sample after the first clustering, recalculate the new average value of each cluster and repeat iterative operation. Assign each sample data to the updated closest average value, and then calculate the average value of the new iterative clustering according to the results. The researchers will specify the number of iterations when implementing the code. Generally the iteration will be carried out many times to ensure that only a few points in the iteration continue to be re-clustered, and finally the expected clustering results will be obtained. Generally speaking, if the number of clusters increases, the number of corresponding iterations will also increase.

4 Research Content

First, we process the mobile phone signaling data and POI interest point data so that they can be applied in the classification method we adopt. Then we adopt supervised classification method and unsupervised classification method to identify and divide the signaling data of Unicom mobile phones. The results obtained by these two methods may be different in terms of accuracy, so how to improve the accuracy of the two classification methods is also a very important content in this paper. Finally, based on the classified results, compare the advantages and disadvantages of the two methods, propose ways for improvement, and analyze and evaluate the classified functional areas in downtown Beijing. The research contents are divided into the following three aspects:

4.1 Processing Original Data

Pre-process the signaling data of Unicom mobile phones and Beijing's POI interest point data in 2018 acquired by means of web crawler tools, unify coordinate systems and units, eliminate unnecessary entries in the research (such as specific names of POI points, enterprise contact number, addresses of POI points, names of cities, postal code of city). Retain the coordinate field and type field of POI points and reduce the total amount of POI data so that the research data can be processed on the computer smoothly. Crop the POI data with the grid data of downtown Beijing, and connect the mobile phone signaling data to the corresponding points on the grid according to the processed coordinates.

4.2 Analyze Mobile Phone Signaling Data Through Random Forest Algorithm

In this paper, random forest algorithm is chosen for supervised classification. Firstly, according to the personnel distribution curve of mobile phone signaling data within 24 h, manually select the points with obvious characteristics in the same class as training samples of this class. A total of 500 points are selected as training sets, and the remaining data are used as prediction data to evaluate algorithm accuracy and output the results. After the above preparations are completed, import the training data set to train the random forest algorithm, and then import the prediction data set. With the trained algorithm, classify and divide the prediction data set and output the final results. The result map obtained by the algorithm is connected with the pre-cropped grid of downtown Beijing to obtain the functional area map of downtown Beijing identified by the random forest method.

In order to improve the accuracy of the algorithm, in this paper, the network density of the processed POI data is analyzed. Firstly, merge the boundaries of the districts in downtown Beijing to generate a complete grid of downtown Beijing. Then, import the data of Beijing's POI points in 2018 and crop it with grid to analyze the POI data in the research area for network density (Fig. 1).

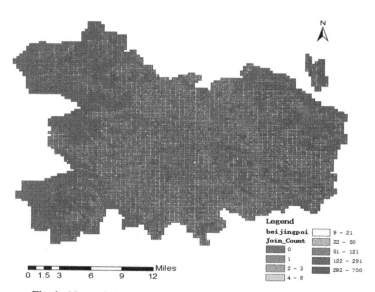

Fig. 1. Network density analysis diagram of POI points in Beijing

According to the number of POI points in each grid through the analysis, we know that the higher the aggregation degree of POI points in a grid, the stronger the functionality of the corresponding city at that point will be, which can help to determine the strength of function in downtown Beijing and other areas. Then, output the results of POI point network density analysis and add them to the characteristics of the random forest to improve the accuracy of the results of random forest algorithm. As can be seen in Fig. 2, in this paper, random forest algorithm is used to divide the functions of downtown Beijing into six categories, namely, green space, residential areas, tourist attractions, office buildings and shopping malls, industrial land and agricultural land. We conclude that each functional area has its own characteristics in terms of data form.

The number of people in green space is relatively small. Population inflow mainly occurs after 7 am when people do morning exercises. During this period, the number of people flowing in and out is roughly the same and the number of population flow decreases after 9 pm when people should have rest. There is basically no population flow in green space after 10 pm. The green spaces are generally scattered around the city because as can be seen in Fig. 2, there are few large area of land reserved for green space in downtown Beijing.

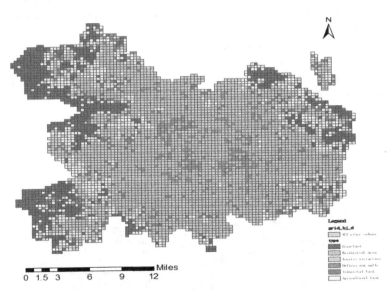

Fig. 2. The functional area of downtown Beijing identified by the random forest algorithm after introducing POI point features (Color figure online)

Population flow in residential areas increases rapidly from 6 a.m., and reaches the maximum at 7 to 8 a.m., which is the rush hour of Beijing. After 9 a.m., the number of mobile phone signaling in residential areas decreases and reaches the minimum at 1 to 5 p.m. because people basically go out to work during the daytime, the number of people in residential areas during this period is the smallest. After 5 p.m., a new wave of population inflow occurs in residential areas since all the people who have finished their work return to their place of residence. After population flow stops, the signaling data generated in residential areas remains unchanged until the rush hour the next day. Such characteristics are consistent with our understanding of the functional area of residential area.

Population inflow occurs in tourist attractions at 7 a.m. Before 7:00, there is only a very small number of mobile phone signaling data records in these areas. Population inflow into these areas significantly increases after 7 a.m., and reaches the maximum at between 12 p.m. and 3 p.m. After 3 p.m., population outflow occurs and the number of people in tourist attractions keeps decreasing until 7 p.m. when mobile phone signaling data reaches the bottom. Such characteristics are in line with our common understanding of the functional area of tourist attraction.

In office buildings and shopping malls, population inflow peaks from 7 am to 9 am, which is the rush hour in the morning. The number of people stabilizes from 9 am to 8 pm when population outflow occurs until 12 pm. After 12 pm, the number of signaling data in these areas reaches the bottom, which means that there are few people there. Office buildings and shopping malls are generally concentrated in relatively developed areas of Beijing, such as Dongcheng District, Xicheng District, Haidian District and Chaoyang District. Such characteristics are in line with our common understanding of the functional areas of office buildings and shopping malls.

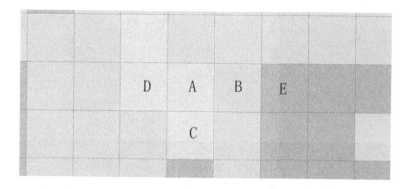

Fig. 3. Sample control diagram. Comparison of random forest classification results with Google earth image

The amount of signaling data of people in industrial land is relatively average within 24 h because the system of 24-h duty shift is adopted in these areas. There are always people there around the clock and there is no large population inflow and outflow. These areas are scattered along the edge of the city, which is in line with our common understanding of industrial land.

The amount of mobile phone signaling data generated in agricultural land is relatively small. Population activity in these land starts at 7 a.m., increases from 9 a.m. and peaks at 3 p.m. After 7 p.m., the number of people gradually decreases, which is in line with our common understanding of agricultural land.

The random forest algorithm has its own accuracy evaluation system, and the accuracy of the random forest algorithm after POI point features are introduced increases from 70% at the beginning to 73%. After the results are obtained, the image data of downtown Beijing generated by Google Map is introduced for further comparison and analysis. The control diagram is as follows (Fig. 3):

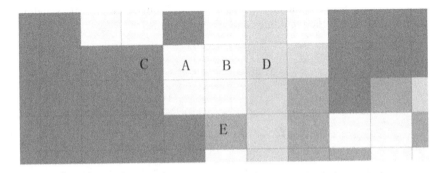

Fig. 4. Sample control diagram. Comparison of random forest classification results with Google earth image

As shown in the figure, Point A, C and D in the results of the random forest classification method represent tourist attractions, corresponding to the Forbidden City and

Zhongshan Park in Google Earth image. Point B represents a residential area, corresponding to the residential areas on the east side of the Forbidden City and Zhongshan Park in Google Earth image. Point E represents office buildings and shopping malls, corresponding to the Wangfujing Pedestrian Street in Google Earth image. The results are consistent (Fig. 4).

The figure shows the area near Wukuangguanshan, Haidian District, Beijing. According to the result of the random forest classification, point A and B represent agricultural land, corresponding to the farmland area in the middle of the Google Earth image. Point D represents a residential area, corresponding to the residential area in the east of the Google Earth image. Point C represents a green space, corresponding to the large forest on the west of the Google Earth image. The results are basically consistent.

4.3 Analyze Mobile Phone Signaling Data Through ISO Clustering Analysis

For unsupervised classification, in this paper, ISO clustering analysis is adopted. First, the author obtains the 24-h mobile phone signaling data on April 17, 2019 and April 20, 2019 on an hourly basis, and export and visualize the mobile phone signaling data of each hour. Figure 5 illustrates the mobile phone signaling data at 0 o'clock. During the calculation, in this paper, the grid of the mobile phone signaling data of the remaining 47 h is adopted.

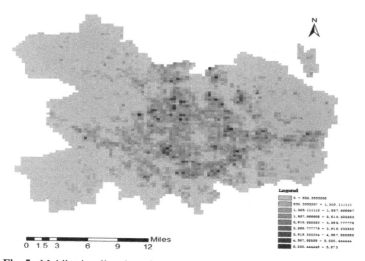

Fig. 5. Mobile signaling data visualization at 0 o'clock in downtown Beijing

The hourly mobile phone signaling raster data set is taken as a separate band, and then the raster data is subjected to band synthesis (a separate raster data set is generated by multiple bands), to generate 24-h band synthesis data of a whole day in downtown Beijing as shown in Fig. 6.

Import this data and realize unsupervised classification through code. Connect the obtained results with the grid of downtown Beijing to generate the functional areas of downtown Beijing by means of unsupervised classification.

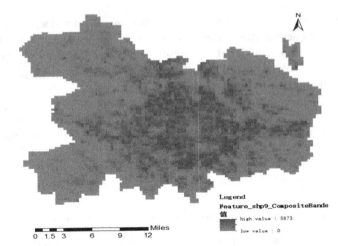

Fig. 6. Mobile signaling data visualization chart of 27 April after band synthesis

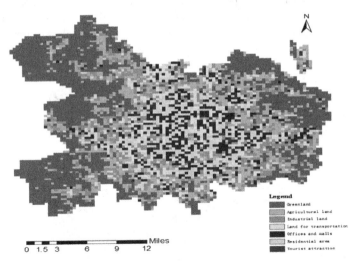

Fig. 7. Functional areas of downtown Beijing calculated by ISO cluster analysis (Color figure online)

As can be seen from Fig. 7, through unsupervised classification, the functional areas of downtown Beijing is divided into seven categories. Then, analyze the results of the unsupervised classification with the remote sensing data and Google earth image data of Beijing and get the specific categories of classification: purple areas are green land, blue and green areas are agricultural land, dark green areas are tourist attraction, dark blue areas are office buildings and shopping malls, orange areas are industrial land, pink areas are residential area and green areas are transportation land.

5 Comparison of Supervised Classification and Unsupervised Classification and Their Advantages and Disadvantages

In terms of supervised classification, firstly, it is necessary to manually select the data with obvious characteristics as training samples according to the characteristics of the data, then select the characteristic parameters and establish corresponding functions to judge the entire sample set. Finally, classify and identify the sample set based on the characteristics of the sample.

To adopt unsupervised classification method, there is no need to provide training samples in advance. The classification and identification are done by computer according to the degree of similarity between sample data without the intervention of researchers. The data categories identified by means of unsupervised classification need to specified with the assistance of traditional data after identification.

The difference between supervised classification and unsupervised classification method mainly lies in whether training samples need to be selected in advance to train the method, and whether researchers need to control how many categories the sample set falls into and the specific categories it falls into. Generally speaking, the effect of unsupervised classification method is not as good as that of supervised classification method.

Supervised classification is generally used by researchers who have a good understanding of the classification areas obtained through the calculation of the research data used. Researchers are required to select training samples with obvious characteristics, then train the algorithm based on the training samples, and classify all the samples with the training samples. Through unsupervised classification, researchers can directly classify the selected data without having a lot of understanding of the selected data, and then judge and classify the data according to its similarity.

In terms of advantages and disadvantages, the main advantage of supervised classification is that it can make full use of the researcher's own understanding of the data to select the training sample set that meets the researcher's needs. In addition, it can screen the training samples many times and let the researchers decide the categories of classification and the number of categories so as to greatly improve classification accuracy and avoid classification errors caused by machines otherwise.

The disadvantage of supervised classification is that researchers will intervene in classification, which requires researchers to spend a lot of time and manpower selecting the required training sample set. Researchers can only identify defined categories, but cannot identify undefined ones through algorithm. Researchers can only classify the data into the category most similar to the defined one. With this method, it is difficult to handle complex classification of multiple types.

The advantage of unsupervised classification is that there is no need for researchers to have deep understanding of the data to be classified because all the classification is completed by machines to avoid errors caused by human intervention. Researchers only need to import the number of required categories, the number of iterations calculated by the algorithm, and the threshold value of classification error. This method works very well if the target needs to be divided into a lot of categories or when researchers do not have deep understanding of the data to be classified.

The main disadvantage of unsupervised classification algorithm is that it needs a lot of research and researchers should have enough common sense to determine the specific category of each result. The common problem is that it is difficult to match the classification results with specific categories. Because of the spectral characteristics of the ground objects used in its classification method, the classification effect is not satisfactory if the spectral characteristics of the two types of ground objects are similar.

6 Conclusion

Import Google Earth Image Map to Analyze Functional Areas in Downtown Beijing
In this paper, mobile phone signaling data and POI interest point data are used and supervised classification and unsupervised classification methods are adopted to identify and divide the functional areas in downtown Beijing. The accuracy of random forest algorithm reaches 73%. In addition, in this paper, Google Earth image data of Beijing is used to further analyze the functional areas of downtown Beijing and conducts research into its residential and commercial areas and public services.

Through the layout of office buildings around residential areas and the mixed layout of different categories of land, in this paper, whether the city layout is reasonable and whether the different categories of land are closely connected are analyzed. In the analysis of the functional areas in eastern and western Beijing, it is found that there are large connected residential areas without any surrounding commercial land, which is not conducive to the sound economic development of eastern and western Beijing. It is necessary to mix residential areas and commercial areas as much as possible to enhance economic vitality in those areas.

Based on the analysis of the layout of industrial land, as well as its surrounding natural conditions and residential area (whether it will affect the normal life of residents), it is found that large industrial land in Fengtai District and Shijingshan District is surrounded by agricultural land, which may lead to environmental pollution. In addition, some industrial land is surrounded by residential areas, which may impact on the daily life of residents there.

Article 12 of Beijing's Urban Master Plan specifies that Beijing will be developed into a world cultural city with cultural confidence, cultural diversity and cultural inclusiveness. Its historical background makes the tourist attractions in its downtown area (such as the Forbidden City, Summer Palace, Tiananmen Square) affect the layout of the surrounding functional areas. Protection of Beijing's cultural characteristics and its urban development should be taken into account simultaneously. While preserving its historical flavor, we should also make innovation. A comprehensive analysis of the distribution of tourist attractions in downtown Beijing and its surrounding functional areas shows that the planning of tourist attractions in Beijing is basically reasonable, but some tourist attractions are surrounded by a large number of residential and commercial areas. The nature of tourist attractions makes such distribution improve the life quality of residents there and increases the population flow in commercial areas, but on the other hand, such distribution may affect the daily life of some residents and lead to noise pollution and traffic congestion.

7 Research Deficiencies and Prospect

Firstly, due to the limitation of data sources, although in this paper, mobile phone signaling data on weekdays and holidays is used, there are still some limitations in the research since only two days' data is used, and the data and experimental data on major holidays and Spring Festival should be different to some extent.

Secondly, in this paper only mobile phone signaling data of China Unicom is used. Although the daily data is detailed and the sample size is large, Unicom users only account for a portion of the total mobile phone users. The way to integrate data collected from various operators should be improved in the research.

Thirdly, in order to improve the accuracy of random forest algorithm, in this paper, the network density of POI data is used, which only represents the functional areas but does not represent the exact categories of classification, so accuracy is improved by only 3%, which is not a significant improvement. The way to adopt new characteristics to improve the accuracy of random forest algorithm needs to be further researched.

Acknowledgements. 国土资源评价与利用湖南省重点实验室开放课题资助 (编号: SYS-MT-201901) "Supported by the Open Topic of Hunan Key Laboratory of Land Resources Evaluation and Utilization.

References

1. Ma, S., Long, Y.: Functional urban area delineations of cities on the Chinese mainland using massive Didi ride-hailing records. Cities **97**, 102532 (2020)
2. Chi, J., Jiao, L., Dong, T., Gu, Y., Ma, Y.: Quantitative identification and visualization of urban functional areas based on POI data. Surv. Mapp. Geogr. Inf. **41**(02), 68–73 (2016)
3. Yu, L., He, X., Liu, J.: Identification of urban functional areas based on spatio-temporal semantic mining. J. Sichuan Univ. (Nat. Sci. Ed.) **56**(02), 246–252 (2019)
4. Jiang, Y., Dong, M., Fan, J., Gao, S., Liu, Y., Ma, X.: Research on identification method of urban functional areas based on POI data. J. Zhejiang Normal Univ. (Nat. Sci. Ed.) **40**(04), 398–405 (2017)
5. Yu, W., Ai, T.: Visualization and analysis of POI points in cyberspace supported by kernel density estimation. Acta Surv. Mapp. **44**(01), 82–90 (2015)
6. Xiao, D., Zhang, X.-Y., Hu, Y.: Identification method of urban function zones based on mobile phone big data. J. Syst. Simul. **31**(11), 2281–2288 (2019)
7. Jiang, G., Hu, R., Shi, L.: Identification of urban functional areas based on call detailed record data. Comput. Appl. **36**(07), 2046–2050 (2016)
8. Chen, Z., Qiao, B., Zhang, J.: J. Earth Inf. Sci. (03) (2018)
9. Jiang, Y., Dong, M., Fan, J., Gao, S., Liu, Y., Ma, X.: J. Zhejiang Normal Univ. (Nat. Sci. Ed.) (04) (2017)
10. Chen, S., Tao, H., Li, X., Zhuoli. J. Geogr. (03) (2016)
11. Cheng, J., Liu, J., Gao, Y.: J. Geosci. (09) (2016)
12. Chai, Y., Shen, Y., Chen, Z.: Human-oriented smart city planning and management based on time-space behavior. Int. Urban Plann. (06) (2014)
13. Application of big data in urban planning: thinking and practice from Beijing institute of urban planning and design. MAO Mingrui. Int. Urban Plann. (06) (2014)

14. Mu, N., Zhang, H., Chen, J., Zhang, L., Dai, H.: Overview of urban application of trajectory data mining. J. Geosci. (10) (2015)
15. Lu, F., Liu, K., Chen, J.: Human mobility in the era of big data. J. Geosci. (05) (2014)
16. Long, Y., Mao, M. Mao, Q., Shen, Z., Zhang, Y.: Refined urban simulation in the era of big data: method data and cases. Hum. Geogr. (03) (2014)
17. Xu, J., Fang, Z., Xiao, S., Yin, L.: Analysis of spatial and temporal differentiation of urban mass mobile phone users. J. Earth Inf. Sci. (02) (2015)
18. Gang, H., Shen, Z., Liu, C., Gu, Y.: A highway speed estimation algorithm based on mobile phone signaling. Electron. World (17) (2017)
19. Xu, H., Xu, J.: Beijing surveying and mapping (06) (2016)
20. Mao, X.: Reconstruction of Zhengzhou's urban function zones from the perspective of functional zoning. Mod. Econ. Inf. (10) (2016)
21. Identification and interaction analysis of urban functional areas based on multi-source data. J. Wuhan Univ. (Inf. Sci. Ed.) (07) (2018)

Spatial Queries Based on Learned Index

Ning Wang and Jianqiu Xu[✉]

Nanjing University of Aeronautics and Astronautics, Nanjing, China
{wn_zcbw,jianqiu}@nuaa.edu.cn

Abstract. With the popularity of location-based services, the scale of spatial data is increasing. Spatial indexes play an important role in spatial databases, and their performance determines the efficiency of data access and query processing. Most of the traditional spatial indexes divide data space or data objects without considering the distribution characteristics of data. In this paper, we design a spatial index structure, named learned Hilbert Model (HM) index. We combine the Hilbert space-filling curve and the two-stage model to build the spatial index. We propose algorithms for point query and range query according to data distribution rules. Experimental results show that the learned HM index can reduce the storage cost by 99% compared with R-tree and Grid Index. Point query efficiency is 40% higher than R-tree and 51% higher than Grid Index. The efficiency of range query is up to 50% higher than R-tree and 57% higher than Grid Index.

Keywords: Spatial index · Learned index · Spatial database · Spatial range query

1 Introduction

With the rapid development of location-based services, query requirements for spatial objects are increasing. Spatial index is a key point for spatial database. Spatial index directly affects the overall performance of spatial databases and geographic information systems. In addition to relying on hardware to improve database system performance, it is necessary to improve the efficiency of spatial index. A number of index structures have been designed and have different applications. Studies have shown that the storage of index structures occupies 55% of the server memory in the commercial databases used today [1]. With the continuous expansion of data scale, the main memory can no longer store the huge index structures. Some indexes need to be stored in external memory, leading to an increasing number of I/O operations [2]. In recent years, big data, machine learning, and artificial intelligence develop rapidly. Data technology has also undergone big changes. Machine learning plays an important role in data quality management and shows its unique advantages in data technology.

For the existing spatial index structures, each has its advantages and disadvantages, but the basic principles are similar. The division method is adopted. The traditional index does not consider the distribution characteristics of the data and assumes the worst data distribution in advance, such that the index is expected to have higher versatility [3]. Spatial range query is one of the most typical spatial queries. Typical indexes such as R-tree [4], Quad-tree [5], and other indexes usually need to traverse through the tree

X. Meng et al. (Eds.): SpatialDI 2020, LNCS 12567, pp. 245–257, 2021.
https://doi.org/10.1007/978-3-030-69873-7_18

structure to find desired data. Excessive datasets result in frequent disk I/O operations. If we can learn the distribution of data and generate a highly adaptive index structure, we can optimize the index, improve the efficiency of data query, and reduce the storage cost.

In this paper, we design an effective spatial index named learned Hilbert Model (HM) index. We use the Hilbert space-filling curve [6] to map multi-dimensional data to one-dimensional data space. Then we construct a two-stage model to learn the distribution of data and predict the position of spatial objects. We propose a point query method and a range query algorithm based on the learned HM index. In the range query algorithm, we divide the query range according to the distribution rule of the Hilbert curve. Using real and synthetic datasets, the experimental results show that the storage cost of learned HM index is 99% less than that of R-tree and Grid Index. The point query efficiency is 40% higher than R-tree and 51% higher than Grid Index. For range queries, the efficiency of the learned HM index is 50% higher than R-tree and 57% higher than Grid Index in the best case.

The rest of this paper is organized as follows. In Sect. 2, we survey the existing studies in spatial index structure, spatial query, and learned index. In Sect. 3, we introduce the learned HM index. In Sect. 4, we design query processing algorithms. The experimental results are shown in Sect. 5. Finally, we conclude the paper and future works in Sect. 6.

2   Related Work

2.1   Spatial Index and Spatial Query

Spatial index is an auxiliary data structure for efficient data access. After years of development, a large number of spatial index structures have emerged. Although a number of index structures and algorithms are designed, they can be divided into space-based and object-based methods [7]. The space-based method divides the geographical space according to the regular or semi-regular way. The spatial objects are divided into corresponding units according to relevant rules, such as the Grid Index [8]. In the object-based method, the spatial objects' division is determined by the position of the objects, such as R-tree.

Spatial query refers to using a spatial index to find spatial objects that meet the query conditions. The efficiency of spatial queries determines whether the database can provide efficient services. Spatial queries can be divided into two types: query based on attribute conditions and query based on geometric conditions. There are several typical spatial queries, such as point query and range query. Point query is the essential query: finding the spatial object at the given location. The range query is to find spatial objects within the given query range. Range query is one of the most popular queries.

2.2   Learned Index

ANN (Artificial Neural Network) [9] is a complex network structure formed by interconnecting many processing units called neurons. Inspired by the working mechanism of biological neural networks, ANN uses mathematical models to simulate neurons' activities. ANN has a single layer or multiple layers. Each layer has multiple neurons. A directed arc connects each neuron with variable weight. ANN "learns" from data by obtaining input values, performing calculations, and repeatedly adjusting weights. ANN has achieved great success in natural language processing, image recognition, and speech recognition.

ANN can learn data distribution and has received extensive attention in recent years. *Kraska et al.* propose a new concept called Learned Index. They regard index as a model. The learned index's central idea is to design a model that can learn the distribution of data and use it to infer the location or existence of data records. They use the Recursive Model Index (RMI) to replace B-tree. The RMI consists of ANN or other linear models. The RMI has significantly better performance than B-tree. It takes up less storage space. Although the learned index of *Kraska et al.* has dramatically improved in query performance and storage cost, the model does not support data insertion. The RMI can only be applied to read-only works. It has significant limitations in the scope of application.

Aiming at the index update problem, *Ding et al.* propose an updatable adaptive learned index called ALEX [10], which solved update problem in the RMI. ALEX can dynamically adjust the RMI structure. ALEX allows each model to independently manage its data and changes the data storage structure to make the insertion and deletion faster. ALEX uses gapped array and packed memory array, that is, array with gaps. The packed memory array provides a gap management mechanism, which can significantly improve the performance of insert operations and reduce data moving steps. The main idea of ALEX is trading space for time. Also, to solve the update problem of learned index, *Gao et al.* propose the Dabble model [11]. They use a clustering algorithm to divide the dataset into K data regions so that the data distribution in each data region is as same as possible. To handle the dynamic insert operation, the Dabble model has a data insertion mechanism based on the middle layer. The middle layer makes the models independent of each other, and makes the Dabble model scalable.

Wang et al. [12] propose the ZM index based on the Z-order curve and multi-layer learning model to apply the learned index on spatial objects. They sort the spatial data with Z-order curve coding and build a multi-layer model. When performing a range query, the ZM index predicts the positions of the upper-left corner and the lower-right corner, and then search for spatial objects in the predicted position range.

3   Learned HM Index

We propose a new index structure named learned HM index. First, we apply the Hilbert curve to sort spatial objects and give each object a unique Hilbert address. Then, we construct the learned HM index by using a two-stage model.

3.1   Hilbert Address

In a one-dimensional data space, data can be sorted in a linear order. The sorted data is monotonous and orderly. Data and its position can correspond one by one. One of the key problems in applying learning models to spatial index is that there is no sequential order in multi-dimensional data. To solve this problem, we apply the Hilbert curve.

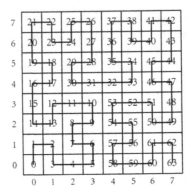

Fig. 1. 3-order Hilbert curve

Hilbert curve is a space-filling curve, which can traverse all points in the unit space to obtain a curve full of space. It maps the multi-dimensional data to one-dimensional space. Each point can get a unique Hilbert address through Hilbert encoding. Here, we take a two-dimensional Hilbert curve as an example. The standard space is a square plane with a side length of 2^n. The Hilbert curve filling this plane is called an n-order Hilbert curve. Figure 1 shows a 3-order Hilbert curve. The generation process of the Hilbert curve is as follows. First, we divide the plane space into four quadrants, and connect the adjacent quadrants with a U-shaped curve. Then, we divide each quadrant into four sub-quadrants, and the sub-quadrants are also connected by U-shaped curves. Repeating this process can produce a curve that fills the entire plane when the degree is infinite.

Typical space-filling curves include Hilbert curve, Z-order curve, and Gray code curve. Hilbert curve has better space aggregation characteristics than other space-filling curves [13]. There is no major abrupt change in the Hilbert curve. It can well preserve the proximity of space objects.

The classical Hilbert coding method is the bitwise algorithm proposed by *Warren M.Lam* and *Jerome M.Shaprio* [14]. For a point (x, y) on a Hilbert curve with order n, the point's Hilbert address is denoted as s. The binary representation of x and y has n bits, and the binary representation of s has $2n$ bits. They use $m[i]$ to represent the value of the i-th bit in the binary sequence m. The values are numbered consecutively from right to left, and the first bit corresponds to the least significant bit. And $m[i:j]$ represents bits i through j of m. The complementation of the corresponding bit sequences is denoted as $m[i:j]'$. The steps of Hilbert coding are shown in Table 1.

To illustrate the key steps of the Hilbert coding, we take the coordinate $(5, 4)$ in the 3-order Hilbert curve as an example. The binary sequence of x and y is represented as 101 and 100. First, based on $x[3] = 1$ and $y[3] = 1$, the value of $s[6:5]$ is 10. The lower bits of x and y are unchanged. Therefore, $x[2] = 0$ and $y[2] = 0$. The table shows that $s[4:3] = 00$, and we need to exchange the lower bits of x and y at the same time. After the exchange, $x[1] = 0$ and $y[1] = 1$. It can be found in the table that $s[2:1] = 01$. The final result of s is 100001. Equivalently the Hilbert address of coordinates $(5, 4)$ is 33.

Table 1. Bitwise operation of Hilbert coding

$x[i]$	$y[i]$	$s[2i:2i-1]$	$x[i-1:1]$	$y[i-1:1]$
0	0	00	$y[i-1:1]$	$x[i-1:1]$
0	1	01	$x[i-1:1]$	$y[i-1:1]$
1	0	11	$y[i-1:1]'$	$x[i-1:1]'$
1	1	10	$x[i-1:1]$	$y[i-1:1]$

3.2 Construction of Two-Stage Model

The main idea of constructing the learned HM index is building a model that can learn the distribution of data and predict the position of the data, or predict a range within which the query data must be included. If the index structure is regarded as a model, machine learning provides many learning models that can learn data distribution and correlation.

As described in Sect. 3.1, the spatial objects are sorted according to their Hilbert address. Given a sorted dataset, a model that can predict the position of key has a good approximation of the cumulative distribution function (CDF). Therefore, it is necessary to model the cumulative distribution function of the data. The position of a given key can be expressed as follows:

$$p = F(key) \times N$$

p is the position of the record. $F(key)$ is the cumulative distribution function of the data, used to calculate the probability that a key is not greater than the given key. N represents the total number of data.

Most machine learning models are not as accurate as traditional index structures in fitting data, and accurate CDF is difficult to learn. However, studies have shown that ANN has a good effect on fitting the overall shape of the CDF [15]. Therefore, a two-stage model containing multiple ANN models is used to construct the learned index.

As shown in Fig. 2, the model contains two stages. The first stage contains one ANN, and the second stage contains multiple ANNs. The entire dataset is divided into multiple subsets. Each ANN model in the second stage is responsible for a specific subset to reduce prediction errors. In the first stage, the query key and its corresponding second-stage model number are used as training data. The trained first-stage model can select model in the next stage according to the input query key. The selected second-stage model predicts the position of the query key. Since each second-stage model is responsible for a specific subset, it can make predictions with smaller errors. Usually, the predicted position may not be the exact position of the query key, but it will be very close to the real position. After the model training is over, we input each key into the model, save the minimum negative error and maximum positive error of each second-stage model by comparing the predicted position with the real position. If a query key exists and its predicted position is denoted as p, the minimum negative error of the corresponding second-stage model is recorded as *min_negative_error*, and the maximum positive error is recorded as *max_positive_error*. It can be determined that the position of the query key must be in the range [*p+min_negative_error, p+max_positive_error*].

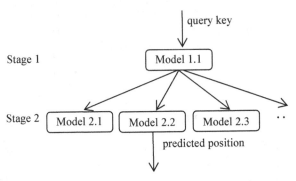

Fig. 2. Two-stage model

The learned HM index is to construct a linear order for spatial objects and learn the distribution from the Hilbert address. Traditional indexes usually need to search in non-leaf nodes to find the appropriate child nodes, while the learned HM index reduces the search process. The result is directly used to locate the position of the data.

4   Query Processing Based on HM Index

We can design effective algorithms to deal with spatial queries using the learned HM index. In this section, we take two-dimensional points as an example and propose methods for point query and range query using the learned HM index. The methods are also suitable for queries in high-dimensional spaces.

4.1   Point Query

Given a query point in a two-dimensional space, its Hilbert address can be calculated and denoted as h. Input h into the learned HM index. The index returns a predicted position p, the minimum negative error and the maximum positive error of the corresponding model denoted as *min_negative_error* and *max_positive_error*. It can be determined that if the query point exists, it must be within the range of [*p+min_negative_error, p+max_positive_error*]. To reduce the search range, we read the record at position p, calculate its Hilbert address, and denote it as hp. If $hp \geq h$, the search range is [*p+min_negative_error, p*]. If $hp < h$, the search range is [*p, p+max_positive_error*], as shown in Fig. 3. By calculating the Hilbert address of record in the search range, binary search is applied to find the real position of the query point.

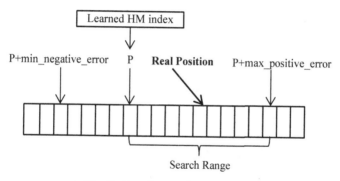

Fig. 3. Search range for point query

4.2 Range Query

To describe the range query process of learned HM index clearly, the following definition is given.

Definition 1. *Hilbert Region*

For a continuous and complete n-order Hilbert curve H, any point $q = (x, y)$ on H satisfies $a \cdot 2^n \leq x \leq a \cdot 2^n + 2^n - 1, b \cdot 2^n \leq y \leq b \cdot 2^n + 2^n - 1$, where a and b are non-negative integers. The smallest square plane S containing H has a side length of 2^n and $\forall p \in H$: $p \in S$. Call S an n-order Hilbert region.

An n-order Hilbert region has the following properties.

(i)   An n-order Hilbert region contains a continuous and complete n-order Hilbert curve.

(ii) The entire n-order Hilbert region can be divided into several i-order Hilbert regions. The number of i-order Hilbert regions is $2^{n-i} \times 2^{n-i}$, where $1 \leq i \leq n$.

(iii) The maximum and minimum values of the Hilbert address appear in the four vertices.

For example, the entire plane is a 3-order Hilbert region, which can be divided into four 2-order Hilbert regions or sixteen 1-order Hilbert regions. No matter how it is divided, in each Hilbert region, the Hilbert curve is completely continuous, and the maximum and minimum values of the Hilbert address in this region must exist in its four vertices.

Figure 4 shows the storage locations of the spatial data sorted according to the Hilbert address. The area shown by the dotted line is a query range. R1, R2, R3, and R4 are the Hilbert regions that intersect the query range. It can be seen that adjacent points in space are not necessarily adjacent in storage, and maybe far apart. In terms of storage, the query data is divided into multiple intervals, which continuously exist in each interval. The points within the query range are contained in all Hilbert regions that intersect the query range.

To decrease the search range, we transform the query range into multiple Hilbert regions. In a standard plane of the n-order Hilbert curve, the steps to divide a query

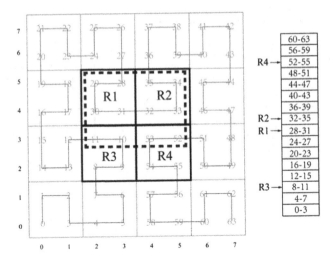

Fig. 4. Partition for query range

window $W = a \times b$ $(a \leq b)$ are as follows. First, we select the short side length of W, and choose the order i which is the order of Hilbert region formed after division. The data space can logically be regarded as a grid composed of i-order Hilbert regions. Then we replace the query range with all i-order Hilbert regions that intersect the query window W.

The query range is transformed into several i-order Hilbert regions. According to the size of each query range, the order of the Hilbert region is different. If the order is too large, the search range will increase fast. If the order is too small, it will result in too many Hilbert regions formed and increase the number of calculations. Selecting the short side length of the query range is to avoid the search range being too large. Taking the order $i = \lfloor \log_2 \frac{1}{2} a \rfloor$ can balance the search range and the number of calculations. Note that the upper bound for i should be set according to the data size and space size. If there is no restriction, i will increase with the query range increasing, which will cause the search range being too large. Figure 4 shows the process of dividing a query range, transforming the query range into four 1-order Hilbert regions. Each Hilbert region corresponds to a storage range, which can effectively exclude many regions that are not within the range.

For a Hilbert region S, the minimum Hilbert address h_start and maximum Hilbert address h_end can be found in its vertices. Input h_start into the learned HM index, it returns the predicted position denoted as p_1, and the minimum negative error of the corresponding model denoted as $min\_negative$. Then input h_end into the learned HM index, we get the predicted position denoted as p_2 and the maximum positive error of the corresponding model denoted as $max\_positive$. The search range of S is $[p_1+min\_negative, p_2+max\_positive]$. Expanding the search range through the error of the model can ensure that the data in the query range is contained in the search range. Perform the same operation on all Hilbert regions and we will get several search range. Within these search ranges, all spatial data in the query range can be found. Algorithm 1 describes the process of range query with learned HM index.

Algorithm 1 Range Query Processing

Input: query range w
Output: range query result r
 1: $data \leftarrow all\_data$
 2: $r \leftarrow \varnothing$
 3: $sub\_w \leftarrow \text{Split}(w)$
 4: $m \leftarrow$ length of sub_w
 5: **for** $i \leftarrow 1$ to m **do**
 6:     $h\_start \leftarrow \text{Hilbert\_start}(sub\_w[i])$
 7:     $h\_end \leftarrow \text{Hilbert\_end}(sub\_w[i])$
 8:     $p\_start \leftarrow \text{Predict}(h\_start)$
 9:     $p\_end \leftarrow \text{Predict}(h\_end)$
10:     **for** $j \leftarrow p\_start$ to p_end **do**
11:         **if** $data[j]$ inside w **then**
12:             $r \leftarrow r \cup data[j]$
13:         **end if**
14:     **end for**
15: **end for**
16: return r

---

5 Experimental Evaluation

In this section, we evaluate the performance of the learned HM index and compare it with R-tree and Grid Index. All the experiments are run on a computer with 64-bit Ubuntu 14.04 system with Intel Core i7-4510U 2.00 GHz CPU, 8G RAM. In the experiments, we use C++ language and a C++ machine learning library mlpack [16] in the extensible database SECONDO [17].

We conduct the experiments using a synthetic dataset and three real-world datasets. The statistics of the datasets is shown in Table 2. POST1, POST2, and POST3 represent data of different scales. POST2 and RANDOM represent evenly and unevenly distributed data with the same data scale. Figure 5 shows the distribution of the real datasets. The distribution of POST2 and POST3 is basically the same, but the data density of POST2 is lower. The index memory size, point query time, and range query time are evaluated on each dataset. In the experiments, the query results of the learned HM index are contrasted with R-tree, which can ensure the correctness of the query results.

The number of second-stage models in the learned HM index is determined by the size of dataset. Each second-stage model corresponds to a subset of 1000. The model used in the second-stage model is a feed-forward neural network with one fully-connected hidden layer. The hidden layer size is 3 neurons. The ReLU (Rectified Linear Unit) function is used as the activation function, and the MSE (mean-square error) function is used as the loss function.

Table 2. Dataset statistics

Dataset	Size	Data sources
RANDOM	200000	Random synthesis
POST1	71003	Restaurants in Beijing
POST2	197489	Companies in Shanghai
POST3	509260	Comprehensive position in Shanghai

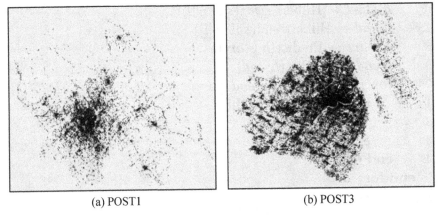

(a) POST1   (b) POST3

Fig. 5. Distribution of dataset

5.1 Index Memory Size

Table 3 shows the index memory size of R-tree, Grid Index, and the learned HM index for different datasets. Compared with the R-tree and Grid Index, the learned HM index reduce memory by 99% on all datasets.

Table 3. Index memory size

Dataset	Index memory size (MB)		
	Learned HM index	R-tree	Grid index
POST1	0.03	7.4	4.2
POST2	0.05	20.6	11.0
RANDOM	0.06	21.5	13.8
POST3	0.11	53.5	26.7

5.2 Point Query

To evaluate the efficiency of the learned HM index, we conduct experiments for point queries on the datasets. For each dataset, we randomly select 1000 points and do point queries. The average query time is measured and shown in Fig. 6. Clearly, the learned HM index has a better performance. Compared with R-tree, the learned HM index is 26% faster on POST1, 40% faster on POST2, 46% faster on RANDOM, and 27% faster on POST3. For the Grid Index, the learned HM index is 35% faster on POST1, 58% faster on POST2, 30% faster on RANDOM, and 51% faster on POST3.

Since the learned HM index has a simpler structure, the position predicted by the model is directly used for data search, which can effectively decrease the search steps and shorten the query time.

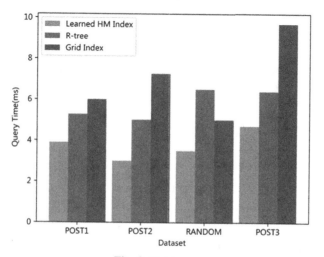

Fig. 6. Point query

5.3 Range Query

Figure 7 shows the results of range query on different datasets by varying the query range size from 0.01% to 10%. Here the query range size represents the ratio of the query range to the MBR (minimum bounding rectangle) of the entire dataset. For each query range size on each dataset, 50 queries are randomly generated to measure the average query time. When generating the query range, we divide the entire data space into uniform grids, randomly select points in each grid as the upper-left point of the query range, randomly generate one side length for the query range, and calculate the other side length to get the entire query range. The query ranges generated in this way can guarantee its randomness and universality.

The learned HM index has a better performance than R-tree on all datasets. On POST1, query time can be reduced by 11% to 37%. On POST2, query time can be reduced by 15% to 50%. On RANDOM, query time can be reduced by 9% to 33%. On POST4, query time can be reduced by 18% to 25%.

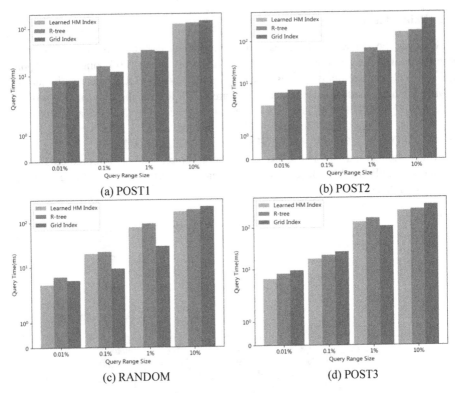

Fig. 7. Range query

Compared with Grid Index, the learned HM index performs better on POST1 and POST2. On POST1, the query time can be reduced by 6% to 25%. On POST2, the query time can be reduced by 22% to 57%. When the query range is 0.1% and 1% on RANDOM, and 1% on POST3, Grid Index performs better. In most scenarios, the learned HM index has a better performance. Grid Index's query time is shorter in some queries, because the Hilbert region after division is too large when using the learned HM index, which causes the search range too large.

6   Conclusion

In this paper, we propose a new index structure named learned HM index, which uses Hilbert space-filling curve and two-stage model to learn the data distribution and predict the positions of spatial objects. We also propose a point query method and a range query algorithm which divides the query range according to the Hilbert curve. Experiments show that learned HM index can significantly reduce the memory size and has good performance in spatial query. It shows that learning models have great potential in spatial data access and query processing.

By making models independent of each other, and adding a corresponding cache area for each second-stage model, insert operation of the learned index have been implemented. We will further evaluate the efficiency of the updateable HM index by experiments. In addition to neural networks, more machine learning models will be tried to improve index performance. More spatial query algorithms based on learned HM index will be designed, such as KNN query, to improve the application scope of the learned index.

Acknowledgement. This work is supported by NSFC under grants 61972198, Natural Science Foundation of Jiangsu Province of China under grants BK20191273 and the Foundation of Graduate Innovation Center in Nanjing University of Areonautics and Astronautics under grants KFJJ20191604.

References

1. Zhang, H., Andersen, D., Pavlo, A., Kaminsky, M., Ma, L., Shen, R.: Reducing the storage overhead of main-memory OLTP databases with hybrid indexes. In: Proceedings of the 2016 International Conference on Management of Data, pp. 1567−1581. ACM (2016)
2. Wu, X., Ni, F., Jiang, S.: Wormhole: a fast ordered index for in-memory data management. In: Proceedings of the Fourteenth EuroSys Conference, pp. 1−16. ACM (2019)
3. Kraska, T., Beutel, A., Chi, E., Dean, J., Polyzotis, N.: The case for learned index structures. In: ACM SIGMOD, pp. 489–504 (2018)
4. Guttman, A.: R-trees: a dynamic index structure for spatial searching. In: ACM SIGMOD, pp. 47–57 (1984)
5. Finkel, R., Bentley, J.L.: Quadtree: a data structure for retrieval on composite keys. Acta Inf. **4**, 1–9 (1974)
6. Sagan, H.: Space-Filling Curves. Springer, New York (1994). https://doi.org/10.1007/978-1-4612-0871-6
7. Rigaux, P., Scholl. M.,Voisard, A.: Spatial databases with application to GIS. In: ACM SIGMOD (2001)
8. Xiao, W., Liao, Y.: Lattice index mechanism in spatial object database system. J. Comput. **017**(010), 736–742 (1994)
9. Kröse, B., Smagt, P.: An introduction to neural networks (1996)
10. Ding, J., Minhas, U. F., Zhang, H., Li, Y., Wang, C., Chandramouli, B.: ALEX: an updatable adaptive learned index (2019)
11. Gao, Y., Ye, J., Yang, N., Gao, X., Chen, G.: Middle layer based scalable learned index scheme. J. Softw. **31**(3), 620–633 (2020)
12. Wang, H., Fu, X., Xu, J., Lu, H.: Learned index for spatial queries. In: MDM. IEEE (2019)
13. Mokbel, M., Aref, W., Kamel, I.: Analysis of multi-dimensional space-filling curves. Geoinformatica **7**(3), 179–209 (2003)
14. Lam, W., Shapiro, J.: A class of fast algorithms for the Peano-Hilbert space-filling curve. In: IEEE International Conference. IEEE (1994)
15. Magdon-Ismail, M., Atiya, A.: Density estimation and random variate generation using multilayer networks. IEEE Trans. Neural Networks **13**(3), 497–520 (2002)
16. Curtin, R., Edel, M., Lozhnikov, M.: mlpack 3: a fast, flexible machine learning library. J. Open Source Softw. **3**(26), 726 (2018)
17. Güting, R., Behr, T., Düntgen, C.: SECONDO: a platform for moving objects database research and for publishing and integrating research implementations. IEEE Data Eng. Bull. **33**(2), 56–63 (2010)

Visualization Science

A Deep Learning Method for Complex Human Activity Recognition Using Virtual Wearable Sensors

Fanyi Xiao, Ling Pei$^{(\boxtimes)}$, Lei Chu, Danping Zou, Wenxian Yu, Yifan Zhu, and Tao Li

Shanghai Key Laboratory of Navigation and Location-Based Services, School of Electrical Information and Electrical Engineering, Shanghai Jiao Tong University, Shanghai, China
ling.pei@sjtu.edu.cn
https://nls.sjtu.edu.cn/

Abstract. Sensor-based human activity recognition (HAR) is now a research hotspot in multiple application areas. With the rise of smart wearable devices equipped with inertial measurement units (IMUs), researchers begin to utilize IMU data for HAR. By employing machine learning algorithms, early IMU-based research for HAR can achieve accurate classification results on traditional classical HAR datasets, containing only simple and repetitive daily activities. However, these datasets rarely display a rich diversity of information in real-scene. In this paper, we propose a novel method based on deep learning for complex HAR in the real-scene. Specially, in the off-line training stage, the AMASS dataset, containing abundant human poses and virtual IMU data, is innovatively adopted for enhancing the variety and diversity. Moreover, a deep convolutional neural network with an unsupervised penalty is proposed to automatically extract the features of AMASS and improve the robustness. In the on-line testing stage, by leveraging advantages of the transfer learning, we obtain the final result by fine-tuning the partial neural network (optimizing the parameters in the fully-connected layers) using the real IMU data. The experimental results show that the proposed method can surprisingly converge in a few iterations and achieve an accuracy of 91.15% on a real IMU dataset, demonstrating the efficiency and effectiveness of the proposed method.

Keywords: Human activity recognition · Inertial measurement units · Deep convolutional neural network · Unsupervised penalty · Transfer learning

1 Introduction

Human activity recognition (HAR) is a research hotspot in the field of computer vision and has broad application prospects in security monitoring, biological

X. Meng et al. (Eds.): SpatialDI 2020, LNCS 12567, pp. 261–270, 2021.
https://doi.org/10.1007/978-3-030-69873-7_19

health, and other fields. Traditional recognition algorithms are mainly based on images or videos [19]. With the emergence of various wearable smart devices embedded with microsensors such as inertial measurement units (IMUs), these devices are highly used in daily life and play an indispensable role in emerging fields that strongly demand HAR such as virtual reality (VR). Therefore, it is a natural way to realize HAR based on wearable devices.

In recent years, HAR based on wearable devices has been conducted deep studies [17], and there exist two general methods. Previous researches use traditional machine learning methods such as Support Vector Machine (SVM) and Random Forest (RF) to receive the recognition result [13,14]. However, these methods need to design features manually, calculate time and frequency domain features based on characteristics of the data. To reduce the computational consumption and compress input data, a further selection of features also needs to be conducted. Due to the longtime design and selection of manual features, it always costs lots using traditional methods of machine learning. With the development of deep learning in recent years, deep neural networks such as Convolutional Neural Network (CNN) [16] or Long Short-Term Memory networks (LSTM) [3] have been widely used for HAR, finishing both feature extraction and activity classification.

Almost all the above methods now can achieve excellent results on specific sensor-based HAR datasets. The widely used public datasets and their main characteristics are shown in Table 1.

Table 1. Widely used public datasets and main characteristics (Acc = accelerometer, Gyro = gyroscope, Mag = magnetometer, Temp = temperature).

Datasets	Sampling rate (Hz)	Sensors	Activities	Subjects
UCI HAR [1]	50	2 (Acc, Gyro)	6	30
WISDM [8]	20	1 (Acc)	7	36
WHARF [2]	32	1 (Acc)	12	17
PAMAP2 [15]	100	4 (Acc, Gyro, Mag, Temp)	18	9

However, all these datasets have defects as follows:

- The most widely used datasets such as UCI HAR [1] contain only simple daily activities, for example, walking, running or jumping, while human behaves much more complex in real life.
- Subjects involved in data collection are always limited, and the same activity tends to be performed similarly, for instance, walking may only include walking at normal speed. However, the same activity can be performed in different styles and may vary with different humans in the real world.
- During data collection, most datasets use only a single IMU, which makes them unsuitable for recognizing more elaborate activities such as stretching

arms or stretching legs. Though other datasets use more than two IMUs, the increase in IMUs also leads to the intrusion to subjects.

To solve problems above, this paper innovatively adopts a pose reconstruction dataset AMASS [10], which is a large collection of motion capture (Mocap) datasets, for HAR. The adoption of this dataset has the following advantages:

1. AMASS contains rich motion types. It includes complex activities such as house cleaning in addition to simple daily activities, making this dataset closer to real life.
2. The containing of multiple mocap datasets in AMASS leads to both richer characteristics in activities and an increase in the number of involved subjects, which is more than 300.
3. Inspired by [6], where virtual IMU data are innovatively used in pose reconstruction, we similarly use virtual IMU for HAR, which greatly reduces the cost of collecting real datasets.

The main contributions of this paper are as follows:

- Adopt a novel pose reconstruction dataset AMASS for HAR and use virtual IMU data in this dataset.
- Use a realistic dataset to fine-tune the model for further reducing the gap between real and virtual data.
- Propose a CNN framework combined with an unsupervised penalty for HAR.

Experimental results show that test result on the realistic dataset is 91.15% after fine-tuning, which demonstrates the feasibility of applying pose recognition datasets and using virtual IMU data for HAR.

2 Dataset Preprocessing Based on the SMPL Model

One major work of this paper is the processing of AMASS, making it suitable for HAR. Since the IMU data in AMASS is virtual, this paper further processes the DIP dataset proposed in [6], which contains real IMU data that can be used to reduce the gap between virtual data and real data.

2.1 SMPL Model

SMPL [9] is a parameterized model of the 3D human body, totally including $N = 6890$ vertices, $K = 23$ joints. Input parameters of this model are shape parameters β, which takes 10 values controlling the shape change of the human body, and pose parameters θ that takes 72 values which define the relative angles of 24 joints (including the root joint) of the human body:

$$M(\beta, \theta) = W(T_P(\beta, \theta), J(\beta), \theta, W), \tag{1}$$

$$T_P(\beta, \theta) = T + B_s(\beta) + B_p(\theta), \tag{2}$$

where T defines a template mesh, to which pose-dependent deformations $B_s(\beta)$ and shape-dependent deformations $B_p(\theta)$ are added. Based on the rotation around the predicted joint locations $J(\beta)$ with smoothing defined by the blend weight matrix W, the resulting mesh is then posed using a standard linear skinning function (LBS).

Using this model, AMASS converts the motion poses of several classical motion capture datasets such as Biomotion [18], from a skeletal form to a more realistic 3D skin model, while the pose parameters are given as a rotation matrix.

2.2   Virtual Data Generation

Though AMASS contains the input parameters of the SMPL model, it does not contain IMU data as original mocap datasets do not provide IMU data. To use AMASS for sensor-based pose reconstruction, [6] confirms the feasibility of synthesizing IMU data and generating corresponding SMPL parameters based on the input of different models.

Based on the rich information provided by AMASS, virtual acceleration data and orientation readings in the rotation matrix can be generated by placing virtual sensors on the SMPL mesh surface. Orientation readings are directly obtained using forward kinematics, while virtual accelerations are calculated via finite differences [6]. The virtual acceleration for time t is defined as:

$$a_t = \frac{p_{t-1} + p_{t+1} - 2 \cdot p_t}{dt^2}, \tag{3}$$

where p_t is the position of a virtual IMU for time t, and d_t is the time interval between two consecutive frames.

2.3   Labeling and Filtering with SMPL Model

Since AMASS contains over 11000 motions, it is necessary to classify these motions into different activities and make true labels. Further, a single motion file in AMASS may consist of several activities, so it is also essential to filter out some motions that affect the balance of the dataset. Activity labeling and data filtering are mainly achieved through three steps.

Posture-Based Labeling. We first classify the whole motions in AMASS into 12 categories based on the superficial descriptions of motions in most classical mocap datasets included in AMASS. Two types of motions are directly removed in this procedure. The first type is motions with little relevance to human daily activities, such as boxing and other martial motions described in Biomotion [18]. The second type refers to some frequently converted motions (e.g. quick transitions between walking, stopping and running). Since the motion duration is generally short in AMASS, frequent motion transitions may conflict with the subsequent sliding window length settings, therefore such motions are also excluded.

Acceleration-Based Filtering. A simple classification of the dataset is implemented in the previous section, while some data are further filtered based on accelerations. Using the accelerations obtained via the sensor on the wrist, a dynamic graph of the acceleration over time can be created. The accelerations at the left wrist for typical walking and running movements are shown in Fig. 1 and Fig. 2.

As can be seen from the comparison of Fig. 1 and Fig. 2, different activities often differ in acceleration characteristics. Therefore, data is further cleaned based on the differences in the characteristics of acceleration (e.g. peaks, variances, etc.).

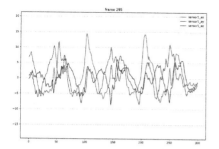
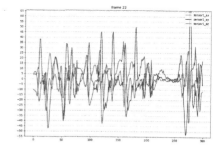

Fig. 1. Wrist accelerations for typical walking movement

Fig. 2. Wrist accelerations for typical running movement

Data Cleaning with SMPL Model. For some activities whose acceleration characteristics are not obvious, such as the stretching of the arms, it almost fails using acceleration features to clean the dataset. However, since AMASS provides SMPL pose parameters in the form of the rotation matrix, it becomes feasible to filter this type of activity adopting visualization with the SMPL model.

After using Unity to build the SMPL model, the motions can be visualized by passing in different SMPL pose parameters. Clapping motion and motion of waving arms are shown in Fig. 3 and Fig. 4 respectively. After visualization of such data, mislabeled motions can be successfully deleted.

However, preprocessed AMASS still suffers the problem of extremely unbalanced activities after processing above, which is mainly caused by unbalanced motions in the original AMASS. To alleviate this problem, interpolation upsampling is adopted in this paper.

3 Deep Learning Algorithm and Fine-Tuning

3.1 Proposed Method

The proposed method includes two stages: the off-line training stage and the on-line testing stage. At the first stage, we firstly employ the AMASS dataset,

Fig. 3. Visualization for clapping movement

Fig. 4. Visualization for waving movement

containing abundant human poses, to enhance the variety and diversity of the real data.

Motivated by the pioneer works [5,7], a deep convolutional neural network (U convolutional layers and S fully-connected layers) with an unsupervised penalty (U deconvolutional layers) is proposed to automatically extract the features of AMASS. Specially, given p-th batch IMU data $\mathbf{X}^{(p)}$ and the related labels $\mathbf{Z}^{(p)}$, the proposed method tries to update the neural network parameters $\Theta = \Theta_0 \cup \Theta_1 = \{\mathbf{W}_k\}_{k=1,2,\cdots,2U+S}$ by minimizing

$$\underset{\Theta}{\operatorname{argmin}} \underbrace{\mathcal{L}_0\left(\Theta_0\right)}_{supervised} + \lambda \underbrace{\mathcal{L}_1\left(\Theta_1\right)}_{unsupervised\ penalty}, \tag{4}$$

where

$$\mathcal{L}_0\left(\Theta_0\right) = \left\| \mathbf{Z}^{(p)} - \varphi_S\left(\mathbf{W}_S\tilde{\mathbf{X}}^{(p)}\right) \right\|_2^2,$$

$$\tilde{\mathbf{X}}^{(p)} = \varphi_U\left(\mathbf{W}_U \cdots \varphi_1\left(\mathbf{W}_1\mathbf{X}^{(p)}\right)\right),$$

$$\mathcal{L}_1\left(\Theta_1\right) = \left\| \mathbf{X}^{(p)} - \varphi_{2U}\left(\mathbf{W}_{2U} \cdots \varphi_{U+1}\left(\mathbf{W}_1\tilde{\mathbf{X}}^{(p)}\right)\right) \right\|_2^2,$$

φ is the activation function of the i-th layer, and λ is the penalty parameter that balances $\mathcal{L}\left(\Theta_0\right)$ and $\mathcal{L}_1\left(\Theta_1\right)$. We use an unsupervised penalty to promote the generalization of the proposed method by considering:

- In our case, by optimizing $\mathcal{L}_0\left(\Theta_0\right)$, we try to represent $\mathbf{X}^{(p)}$ of high-dimension by the latent layer of low-dimension ($\tilde{\mathbf{X}}^{(p)}$). Such an operation, considering the low dimensionality of the IMU data, is helpful for the key feature extraction.
- The unsupervised penalty in (4) itself is a denoising autoencoder [7] that can help denoising the AMASS dataset, enhancing the robustness of the proposed method.
- It has been shown in previous studies that learning multi-task (i.e., $\mathcal{L}_0\left(\Theta_0\right)$ and $\mathcal{L}_1\left(\Theta_0\right)$) jointly can improve the generalization error bounds [4,11].

3.2 Fine-Tuning with Real IMU

Since IMU data in AMASS is virtually generated via the SMPL model and virtual sensors, while the IMU data in the real world tends to be affected by environmental noise, electromagnetic waves, etc. Therefore, certain differences exist between virtual and real data. To eliminate the gap, this paper uses the DIP dataset with real IMU data provided in [6] for fine-tuning. Data processing of DIP is similar to AMASS, except the fact that DIP only contains 5 activities, namely "computer works", "walking", "jumping", "stretching arms" and "stretching legs". Meanwhile, DIP has rather balanced activity categories, therefore up-sampling is not performed on DIP.

Following the off-line training stage in Sect. 3.1, at the on-line testing stage, by leveraging advantages of the transfer learning, we obtain the final result by fine-tuning the parameters in the fully-connected layers with the real IMU data.

4 Test Verification

This paper innovatively adopts a pose reconstruction dataset AMASS with virtual IMU data for HAR and proposes a new CNN framework with an unsupervised penalty. We design several comparative experiments, to prove the feasibility of using pose reconstruction dataset for HAR.

To further verify the rationality of the method proposed in this paper, both classical machine learning algorithms and deep learning algorithms are tested on AMASS and DIP. Taking the sequence length in AMASS into consideration, this paper finally adopts RF and DeepConvLSTM [12] algorithms for comparisons. For RF we directly input the processed data for classification, while we adopt the original DeepConvLSTM architecture for comparison.

4.1 Experimental Design

Three groups of comparative experiments based on different datasets are designed. Experiment 1 conducts training and testing on AMASS, using all three algorithms. The ratio of the training set to the test set is 7:3. Experiment 2 conducts training and testing on DIP and adopts all three algorithms similar to experiment 1. Experiment 3 is trained on the AMASS training set, fine-tuned on the DIP training set and finally tested on the DIP test set. Only our proposed method and DeepConvLSTM are involved in experiment 3.

Considering that some activities cannot be identified using only one IMU, three IMUs located at the left wrist, the right thigh, and the head are selected in this paper. The total input data have features in 36 dimensions, including three-axis acceleration and rotation matrix. Since the sampling rates of AMASS and DIP are both 60 Hz, a sliding window with 60 frames (i.e. 1 s) length is selected, while the degree of overlapping is set as 50%.

4.2 Evaluation Criteria

Commonly used evaluation criteria in HAR are accuracy, recall, F1-score and Area Under the Curve (AUC), among which accuracy and F1-score are most commonly used. Therefore, we also adopt accuracy and F1-score as the performance measures:

$$Accuracy = \frac{\sum_{cn=1}^{CN} TP_{cn} + \sum_{cn=1}^{CN} TN_{cn}}{\sum_{cn=1}^{CN} TP_{cn} + \sum_{cn=1}^{CN} TN_{cn} + \sum_{cn=1}^{CN} FP_{cn} + \sum_{cn=1}^{CN} FN_{cn}}, \quad (5)$$

$$F1\text{-}score = \frac{2\sum_{cn=1}^{CN} TP_{cn}}{2\sum_{cn=1}^{CN} TP_{cn} + \sum_{cn=1}^{CN} FP_{cn} + \sum_{cn=1}^{CN} FN_{cn}}, \quad (6)$$

where CN denotes the class number. Variables TP_{cn}, FP_{cn}, TN_{cn}, FN_{cn} are the true positives, false positives, true negatives and false negatives of the class cn, respectively.

4.3 Experimental Results and Analysis

Table 2 illustrates all results in three experiments. From the results on the AMASS dataset in Table 2, we can see that all three algorithms can achieve accuracy over 70%, despite the fact that IMU data in AMASS is virtual and the containing of complex activities composed of several motions. Results on the DIP dataset in Table 2 corresponds to the results of experiment 2, comparing three algorithms on a realistic IMU dataset DIP. We can see that the proposed method outperforms DeepConvLSTM and RF on both AMASS and DIP, which strongly illustrates the rationality of the deep learning algorithm proposed in this paper.

Table 2. Experimental results

Dataset	Methods and results					
	Proposed method		DeepConvLSTM		RF	
	Acc	F1-score	Acc	F1-score	Acc	F1-score
AMASS	**87.46%**	**86.50%**	73.03%	72.43%	75.01%	70.00%
DIP	**89.08%**	**89.16%**	78.33%	79.31%	77.25%	75.96%
AMASS & DIP	**91.15%**	**91.21%**	84.80%	85.12%	\	\

Notice that the classification result on DIP is not as good as the classification result of DeepConvLSTM and RF on classical HAR datasets. The main reason is that although DIP only contains 5 activities, similar to AMASS, each activity may be composed of a variety of motions, such as activity stretching legs which includes two motions, leg raising, and stepping. Activities with multiple motions greatly increase the difficulty of classification.

To confirm gaps between virtual IMU data and real IMU data, we additionally use the proposed network trained on AMASS to finish the classification task on DIP, an unsurprising result of accuracy less than 50% is obtained. While the network trained based on AMASS and fine-tuned on the DIP training set achieves the best performance on the DIP test set, both for the proposed method and DeepConvLSTM. The results confirm that fine-tuning indeed eliminates the gap between the virtual IMU and the real IMU to some extent.

We also show the confusion matrix figures of the proposed method in experiment 2 experiment 3. As Fig. 5 and Fig. 6 show, fine-tuning effectively improves the classification results of some categories in DIP, which is mainly caused by richer motions in AMASS that make it easier to distinguish some confusing activities. Another interesting thing to be noticed is that fine-tuning can achieve rather excellent results within 20 epochs. This also provides a way for future research, that is, training on large-scale virtual IMU datasets, only need for a small scale of datasets with real IMU data for fine-tuning, which will reduce the cost of collecting real data.

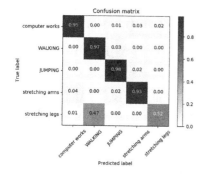

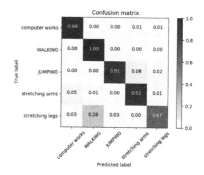

Fig. 5. Confusion matrix of DIP in experiment 2

Fig. 6. Confusion matrix of DIP in experiment 3

5   Conclusion

This paper innovatively adopts a pose reconstruction dataset AMASS for HAR for the problem of simple daily activities and limited subjects in classical datasets. At the same time, a pose reconstruction dataset DIP with real IMU data is used for fine-tuning, to reduce the gap between virtual IMU data and real IMU data. Future work can focus on the most suitable IMU configurations through more detailed experiments.

Acknowledgment. This work was supported by the National Nature Science Foundation of China (NSFC) under Grant 61873163, Equipment Pre-Research Field Foundation under Grant 61405180205, Grant 61405180104.

References

1. Anguita, D., Ghio, A., Oneto, L., Parra, X., Reyes-Ortiz, J.L.: A public domain dataset for human activity recognition using smartphones. In: ESANN (2013)
2. Bruno, B., Mastrogiovanni, F., Sgorbissa, A.: Wearable inertial sensors: applications, challenges, and public test benches. IEEE Robot. Autom. Mag. **22**(3), 116–124 (2015)
3. Chevalier, G.: LSTMs for human activity recognition (2016)
4. Chu, L., Li, H., Qiu, R.C.: LEMO: learn to equalize for MIMO-OFDM systems with low-resolution ADCs. arXiv preprint arXiv:1905.06329 (2019)
5. Erhan, D., Bengio, Y., Courville, A.C., Manzagol, P.A., Bengio, S.: Why does unsupervised pre-training help deep learning? J. Mach. Learn. Res. **11**(3), 625–660 (2010)
6. Huang, Y., Kaufmann, M., Aksan, E., Black, M.J., Hilliges, O., Pons-Moll, G.: Deep inertial poser: learning to reconstruct human pose from sparse inertial measurements in real time. ACM Trans. Graph. (TOG) **37**(6), 1–15 (2019). Article no. 185
7. Lecun, Y., Bengio, Y., Hinton, G.: Deep learning. Nature **521**(7553), 436 (2015)
8. Lockhart, J.W., Weiss, G.M., Xue, J.C., Gallagher, S.T., Grosner, A.B., Pulickal, T.T.: Design considerations for the WISDM smart phone-based sensor mining architecture. In: Proceedings of the Fifth International Workshop on Knowledge Discovery from Sensor Data, pp. 25–33. ACM (2011)
9. Loper, M., Mahmood, N., Romero, J., Pons-Moll, G., Black, M.J.: SMPL: a skinned multi-person linear model. ACM Trans. Graph. (TOG) **34**(6), 1–16 (2015). Article no. 248
10. Mahmood, N., Ghorbani, N., Troje, N.F., Pons-Moll, G., Black, M.J.: AMASS: archive of motion capture as surface shapes. arXiv preprint arXiv:1904.03278 (2019)
11. Maurer, A., Pontil, M.: Excess risk bounds for multitask learning with trace norm regularization. J. Mach. Learn. Res. **30**, 55–76 (2013)
12. Ordóñez, F., Roggen, D.: Deep convolutional and LSTM recurrent neural networks for multimodal wearable activity recognition. Sensors **16**(1), 115 (2016)
13. Pei, L., et al.: Human behavior cognition using smartphone sensors. Sensors **13**(2), 1402–1424 (2013)
14. Pei, L., Liu, J., Guinness, R., Chen, Y., Kuusniemi, H., Chen, R.: Using LS-SVM based motion recognition for smartphone indoor wireless positioning. Sensors **12**(5), 6155–6175 (2012)
15. Reiss, A., Stricker, D.: Introducing a new benchmarked dataset for activity monitoring. In: 2012 16th International Symposium on Wearable Computers, pp. 108–109. IEEE (2012)
16. Ronao, C.A., Cho, S.-B.: Human activity recognition with smartphone sensors using deep learning neural networks. Expert Syst. Appl. **59**, 235–244 (2016)
17. Sousa Lima, W., Souto, E., El-Khatib, K., Jalali, R., Gama, J.: Human activity recognition using inertial sensors in a smartphone: an overview. Sensors **19**(14), 3213 (2019)
18. Troje, N.F.: Decomposing biological motion: a framework for analysis and synthesis of human gait patterns. J. Vis. **2**(5), 2 (2002)
19. Zhang, S., Wei, Z., Nie, J., Huang, L., Wang, S., Li, Z.: A review on human activity recognition using vision-based method. J. Healthc. Eng. **2017**, 1–31 (2017)

A Geometric Consistency Model of Virtual Camera for Vision-Based SLAM Simulation

Yifan Zhu, Ling Pei(✉), Danping Zou, Wenxian Yu, Tao Li, Qi Wu, and Songpengcheng Xia

Shanghai Key Laboratory of Navigation and Location-based Services, School of Electronic Information and Electrical Engineering, Shanghai Jiao Tong University, Shanghai, China
ling.pei@sjtu.edu.cn
http://nls.sjtu.edu.cn/

Abstract. The performance of vision-based simultaneous localization and mapping (VSLAM) algorithms is affected by physical space, environment variables, and other factors, which require massive verifications in diverse scenarios in the real world. However, collecting visual data for VSLAM algorithms in the real world is an expensive and time-consuming process. With the development of rendering technology, it is possible to directly generate data sets using synthetic images with a virtual camera using computer simulation. In order to simulate realistic images with a virtual camera, precise modeling of the geometric characteristics of the vision sensor must be addressed. In this paper, we propose a geometric consistency model considering both the projection characteristics and the lens distortion of a camera. We also provide an efficient implement of the proposed geometric consistency model, which can be used to generate data sets or evaluate algorithms.

Keywords: Synthetic image · SLAM · Camera model

1 Introduction

The vision-based simultaneous localization and mapping (VSLAM) is becoming an increasingly important topic in both robotics and computer vision. The purpose of VSLAM is to recover the camera trajectory and reconstruct the map based on environmental features (points, lines) recorded by the vision sensor [9,18]. The development and evaluation of the VSLAM algorithm need a lot of real data verification which makes generating data sets in different scenarios become one frequently mentioned topic in VSLAM research. A standard VSLAM data set includes images taken by a camera and the pose of the camera at the same time. It is often necessary to lay expensive systems such as Vicon [7] in indoor environments and high accurate GNSS/IMU georeference system for outdoors to obtain accurate pose of a camera in the real world, which is unaffordable to many researchers.

X. Meng et al. (Eds.): SpatialDI 2020, LNCS 12567, pp. 271–280, 2021.
https://doi.org/10.1007/978-3-030-69873-7_20

To address this problem, many open-source data sets are published nearly every year covering a variety of platforms (vehicle, mobile robot, unmanned aerial vehicle, etc.) and environments (urban, indoor, underwater, etc.) [2,17]. However, the open-source data sets are not well-qualified due to several reasons. Firstly, most of the published data sets are generated discretely, which do not allow users to freely re-sample the data. Secondly, many data sets are collected for a specific scenario, which means researchers cannot change environmental parameters later. Thirdly, data sets are usually collected using a particular camera and platform, which makes it difficult for users to test other cameras or platforms in the same environment.

An alternative solution is to generate data sets directly using the renderer running on a computer [14,16]. The modern game engine (such as Unreal Engine4, Unity) achieves impressive improvement to simulate the physical world and allows real-time rendering of photo-realistic RGB images. Several virtual cameras (UnrealCV, Microsoft AirSim [10,12]) based on the modern game engine have been published to generate data sets for different purposes such as deep learning, reinforcement learning, and autonomous driving. These virtual cameras are intent on the algorithms such as object identification, scene segmentation, and perception which are not sensitive to the geometric characteristics of vision sensors. Therefore, these characteristics are ignored in the above virtual cameras. However, the geometric characteristics are essential in VSLAM algorithms for accurate pose estimation. In order to address this issue, we propose a model that can keep the geometric consistency between images photo by a real camera and synthetic images generated by a virtual camera. The contributions of this paper are given as follows:

- We propose a geometric consistency model to simulate the projection characteristics and lens distortion of a vision sensor on a standard computer.
- An open-source implement of a virtual camera with the proposed geometric consistency model is developed as a plugin of Unreal Engine4 which can be found at https://github.com/DrifterFun/Geometric-Consistency-Model.

2   Related Work

The attention on a solution of using computers to generate images and compose VSLAM data sets has been growing considerably. Gazebo [6] is an initial work to support 3D dynamic multi-robot simulation, which allows users to define different types of cameras and robots in a scene. However, Gazebo has a low performance when applied to large-scale scenes with rich visual features. UnrealCV [10] develop a virtual camera based on the Unreal Engine4 (UE4) that guarantees an impressive performance in generating synthetic images. It is capable of handling large-scale scene and can arbitrarily set the illumination intensity, the position of light source, the placement of objects in the scene. Besides, UnrealCV also allows real-time rendering realistic synthetic images owing to advanced UE4 rendering architecture. Similarly based on UE4, AirSim [12] recently proposed provides a number of complex visually rich environments and a weather system.

There is another example in the autonomous driving area named Carla [3]. It utilizes virtual cameras to generate synthetic images for training reinforcement learning algorithms. Besides, UnrealNavigation [1] is an integrated platform specially designed for testing VSLAM algorithms, which embeds Robotic Operating System (ROS) in its framework. In general, virtual cameras have evolved from relying on a single renderer to using modern game engine architectures. At the same time, the use of synthetic images to test and study SLAM algorithms has also attracted accumulating attention.

3 Geometric Consistency Model

The proposed geometric consistency model simulates a real camera in the virtual environment with a pinhole camera model and lens distortion model on a standard computer. The pinhole camera model describes the projection characteristics of a vision sensor and the lens distortion model describes the radial and tangential distortion of a vision sensor in the virtual world.

3.1 Pinhole Camera Model

For the convenience of description, we plot the pinhole camera model in Fig. 1a. Three coordinate systems are used in this paper: camera coordinate system (CCS), retinal coordinate system (RCS), and pixel coordinate system (PCS). The CCS is a three-dimensional coordinate system centered on the camera focus, represented as $X_c - Y_c - Z_c$ in Fig. 1a. The RCS is a two-dimensional coordinate system defined on the camera's image plane, represented as $x - y$ in Fig.1a. The PCS is defined on the image with its coordinate origin located at the upper left corner of the image, shown as $u - v$ in Fig. 1a.

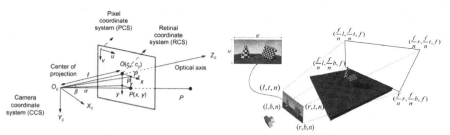

(a) Diagram of coordinate system.   (b) Limited viewing frustum of pinhole camera model in virtual world.

Fig. 1. Schematic diagram of Pinhole camera model.

The pinhole camera model can be mathematically described as a linear transformation in homogeneous coordinates, which can be written as formula (1).

$$\begin{bmatrix} u \\ v \\ 1 \end{bmatrix} = \begin{bmatrix} \alpha f & 0 & c_x \\ 0 & \beta f & c_y \\ 0 & 0 & 1 \end{bmatrix} \begin{bmatrix} x_c/z_c \\ y_c/z_c \\ 1 \end{bmatrix} \tag{1}$$

Where f is the focal length. α and β are the scale factor in u and v direction. c_x and c_y are the offset in the u and v direction. $[x_c, y_c, z_c]^T$ is the point in CCS and $[u, v]^T$ is the point in PCS.

In order to simulate the physical imaging process of the pinhole camera model in a virtual world with limited computation resources, the rendering engine restricts the imaging area to a limited viewing frustum (as showing in Fig. 1b). The scene laid inside viewing frustum is projected on the near plane, and the limited viewing frustum is decided by six parameters: near(n), left(l), right(r), top(t), bottom(b) and far(f). The image projected onto the near plane is discretized to a specified resolution (WxH) to generate the desired image.

To maintain the geometrical consistency between the virtual camera with a rendering engine and the real vision sensor, the conversion formula of the pinhole camera model parameters and the rendering engine parameters need to be established. The pinhole camera model has a total of five parameters (camera intrinsics) to describe the process to project a point from the CCS to the PCS, while six parameters are required in the rendering engine. This is because the viewing frustum is not limited in the pinhole camera model, so far(f) is equal to infinity. When a specific vision sensor is determined, other parameters of the rendering engine can be uniquely determined. To make our geometric consistency model suitable for all vision sensors, we have introduced the conversion formulas between the camera model parameters and the rendering engine parameters. Meanwhile, our open-source implement can adaptively generate a camera model that meets the requirements when users only give pinhole camera model parameters.

When the resolution of an image is given (WxH), the conversion formula for camera intrinsics and the parameters of a rendering engine can be obtained by formula (2).

$$\begin{cases} f = near \\ \alpha = W/(right - left) \\ \beta = H/(top - bottom) \\ c_x = (right + left)W/2(right - left) + W/2 \\ c_y = (top + bottom)H/2(top - bottom) + H/2 \end{cases} \qquad \begin{cases} near = f \\ right = c_x/\alpha \\ left = (c_x - W)/\alpha \\ top = c_y/\beta \\ bottom = (c_y - H)/\beta \end{cases} \tag{2}$$

3.2   Lens Distortion Model

In reality, a camera usually contains a set of lenses in order to get a clear image. The addition of the lens makes the projection characteristics of a vision sensor slightly deviate from the pinhole camera model because it changes the propagation path of light. The phenomenon that image data no longer conforms to the pinhole camera model because of a lens is called "lens-distortion".

The academic community models the phenomenon of "lens-distortion" as two kinds of components: "radial distortion" and "tangential distortion" [11]. The radial distortion model is used to model the distortion caused by the lens shape, and the tangential distortion model is used to describe the distortion introduced by the lens that cannot be completely parallel to the imaging plane. The camera distortion model can be expressed by the formula (3).

$$
\begin{cases}
x_d = x + \underbrace{x(k_1 r^2 + k_2 r^4 + k_3 r^6)}_{\text{Radial Distortion}} + \underbrace{2p_1 xy + p_2(r^2 + 2x^2)}_{} \\
y_d = y + y(k_1 r^2 + k_2 r^4 + k_3 r^6) + \underbrace{p_1(r^2 + 2y^2) + 2p_2 xy}_{\text{Tangential Distortion}}
\end{cases}
\tag{3}
$$

Where $[x, y]^T$ is the point without distortion in RCS, and $[x_d, y_d]^T$ is the point with distortion in RCS. k_1, k_2, k_3 are the parameters of radial distortion and p_1, p_2 are the parameters of tangential distortion. r, which indicates the distance between $[x, y]^T$ and the origin of RCS, could be expressed by formula (4).

$$
r^2 = x^2 + y^2
\tag{4}
$$

In order to simulate the distortion of the lens, we need to change the projection rule of the 3D point in the viewing frustum slightly. In the ordinary pinhole camera model, 3D points inside the viewing frustum are projected onto the imaging plane along the line passing through the focus. When taking lens distortion into account, we need to adjust the propagation path of light in the viewing frustum according to the lens distortion model. We design an efficient method to achieve the above process: the scene is first rendered into an undistorted image using an unmodified graphics pipeline based on a pinhole camera model and then a fragment shader can transform an undistorted image into a distorted image in the postprocessing step.

The postprocessing step requires the computation of undistorted pixel coordinates $[x, y]^T$ from distorted pixel coordinates $[x_d, y_d]^T$, i.e. the inverse of formula (3). Drap [4] proposed an exact inversion for radial distortion. But this inversion is not fully solved for the full set of parameters of the original standard model. In the literature, there are several solutions to overcome this problem (Weng et al. [15]) such as approximation, iteration, and look-up table. The approximation method is more suitable for our application because the iterative method is time-consuming and the look-up table method is memory-consuming.

According to the existing approximation methods, we apply the ideas by Heikkilä [5] to invert formula (3) using an approximation based on Taylor series. While his model only supports radial distortion parameters k_1, k_2 and tangential distortion parameters p_1, p_2. We extend the approximation method to support k_3, which is sufficient in practice

$$
\begin{cases}
x' = x_d - b(ax_d + 2p_1 x_d y_d + p_2(r^2 + 2x_d^2)) \\
y' = y_d - b(ay_d + p_1(r^2 + 2y_d^2) + 2p_2 x_d y_d) \\
a = k_1 r^2 + k_2 r^4 + k_3 r^6 \\
b = 1/(4k_1 r^2 + 6k_2 r^4 + 8k_3 r^6 + 8p_1 y_d + 8p_2 x_d + 1)
\end{cases}
\tag{5}
$$

Where $[x^{'}, y^{'}]^T$ is the estimated point without distortion in RCS, and $[x_d, y_d]^T$ is the point with distortion in RCS. $r^2 = x_d^2 + y_d^2$ indicate the euclidean distance between $[x_d, y_d]^T$ and the origin of RCS.

4   Experiments and Results

4.1   Comparison Between Real Images and Synthetic Images

In this experiment, the accuracy of the geometric consistency model was evaluated with the comparison of real images and synthetic images. We tested our geometric consistency model with the calibration part of a VSLAM data set proposed by J. Sturm [13]. First, we build our geometric consistency model in the UE4 environment with the camera intrinsic matrix and lens distortion parameters given in the dataset. Then we generated synthetic images of checkerboard at the same poses as the real images. After that, the intersections of checkerboard were detected in both real images and synthetic images. Finally, we compared the differences between the two sets of intersections to measure the performance of the proposed geometric consistency model. We plot the pipeline of evaluation in Fig. 2. The intersection point in the calibration board is easy to be detected automatically in the image, so it is chosen for evaluation. Besides, this method is also widely used in the field of computer vision.

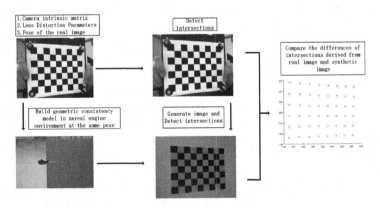

Fig. 2. The pipeline of evaluation. The real image above is taken from J. Sturm's open-source SLAM dataset [13].

We took 15 real images in total and generated their corresponding synthetic images. For each generated image, we extracted 48 intersections for analysis, which means that we had 720 intersections pairs for evaluation. We give the performance of our geometric consistency model in terms of E_{pixels}, E_{norm}, which are defined as formula (6). E_{pixels} represents the pixel distance error in PCS and E_{norm} represents the norm distance error in RCS.

$$\begin{cases} E_{pixels} \equiv \frac{1}{N} \sum_{i=0}^{N} \sqrt{\left(x_i^S - x_i^R\right)^2 + \left(y_i^S - y_i^R\right)^2} \\ E_{norm} \equiv \frac{1}{N} \sum_{i=0}^{N} \sqrt{\frac{1}{f_x^2}\left(x_i^S - x_i^R\right)^2 + \frac{1}{f_y^2}\left(y_i^S - y_i^R\right)^2} \end{cases} \quad (6)$$

where N represents the number of corresponding intersections. f_x and f_y are the camera intrinsics which can be obtained from dataset. $[x_i^S, y_i^S]^T$ and $[x_i^R, y_i^R]^T$ represent the coordinates (PCS) of ith pixel in the synthetic image and the real image, respectively.

Table 1. Pixel distance error between real images and synthetic images.

	Microsoft Kinect AirSim's Model [12]	Proposed model without Lens Distortion Model	Proposed model with Lens Distortion Model
E_{norm}	0.0352	0.0034	0.0018
E_{pixels}	18.18	1.78	0.95

In Table 1, the proposed model can attain sub-pixel accuracy on the "Microsoft Kinect" camera (E_{pixels}=0.95) and performs better than the Air-Sim's model. Experimental results also show that E_{norm}=0.0018, which means that the proposed model can attain sub-pixel accuracy when the parameters of a camera meet the demand where $\frac{1}{E_{norm}} > \max\{f_x, f_y\}$.

4.2 VSLAM Dataset Generation

In this experiment, we show the help of the virtual camera with the proposed geometric consistency model for VSLAM. We used the data generated by the virtual camera to explore the impact of the estimation of camera lens distortion model parameters on the VSLAM algorithm. The advantage of using the virtual camera is that we can flexibly and freely set the camera's precise pose and parameters in a virtual environment to generate data sets, which is time-consuming to achieve in the real world.

Fig. 3. Synthetic image data sets generated by proposed virtual camera

We build the proposed virtual camera in the "Modular Neighborhood Pack" which is an open-source environment in the UE4 market. As shown in Fig. 3, the coordinates of our virtual camera have been marked in the figure and the X-axis (red) direction is the orientation of the camera. In order to get the absolute scale of the world, we used the virtual camera to generate a data set of the stereo camera. For the convenience of analysis, we assume that the virtual stereo camera moves counterclockwise at a speed of 2.4 m/s along the red line in Fig. 3. At the same time, the virtual stereo camera generates 25 frames of image data and corresponding poses per second. In theory, our virtual camera could move along any curve in the scene. We choose a relatively simple curve (red line in Fig. 3), for analyzing the positioning error more easily.

Table 2. The geometric parameters of two stereo camera datasets

Geometric parameters	f_x	f_y	c_x	c_y	k_1	k_2	k_3	p_1	p_2	Baseline	Resolution
D1-L-Camera	640.0	640.0	640.0	360.0	−0.1	−0.05	−0.025	0.0	0.0	1.205 m	720P
D1-R-Camera	640.0	640.0	640.0	360.0	−0.1	−0.05	−0.025	0.0	0.0	1.205 m	720P
D2-L-Camera	640.0	640.0	640.0	360.0	0.0	0.0	0.0	−0.1	−0.1	1.205 m	720P
D2-R-Camera	640.0	640.0	640.0	360.0	0.0	0.0	0.0	−0.1	−0.1	1.205 m	720P

As shown in Fig. 3, we generated two stereo camera data sets using different camera parameters to explore the effects of radial distortion and tangential distortion parameters estimation on the SLAM algorithm. The geometric parameters of two stereo cameras are listed in Table 2. It is worth noting that any of the parameters in Table 2 can be set through the proposed model and two data

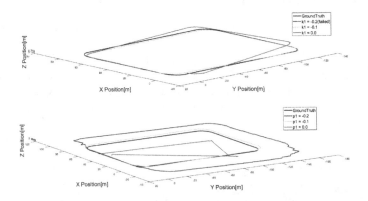

Fig. 4. The results of running ORB-SLAM with different distortion parameters in Dataset1 & Dataset2 Radial distortion is evaluated with Dataset1 and the parameters are stepped through a range of $k_1 = \{-0.2, -0.1, 0.0\}$ and $k_2 = k_1/2$, $k_3 = k_1/4$, $p_1 = p_2 = 0$. Tangential distortion is evaluated with Dataset2 and the parameters are stepped through a range of $p_1 = \{-0.2, -0.1, 0.0\}$ and $p_2 = p_1$, $k_1 = k_2 = k_3 = 0$. (Color figure online)

sets with a total size of 25.8 GB were generated in 86 mins on a computer (Intel I5-8500 & NVIDIA GTX-1660Ti & 8G DDR4-RAM & 512G SSD).

Table 3. Position error under different distortion parameters

Dataset-1						Dataset-2					
k_1	k_2	k_3	p_1	p_2	RMSE[m]	k_1	k_2	k_3	p_1	p_2	RMSE[m]
-0.2	-0.1	-0.05	0.0	0.0	Failed	0.0	0.0	0.0	-0.2	-0.2	24.38
-0.15	-0.075	-0.0375	0.0	0.0	5.33	0.0	0.0	0.0	-0.15	-0.15	11.02
-0.1	**-0.05**	**-0.025**	**0.0**	**0.0**	**3.16**	**0.0**	**0.0**	**0.0**	**-0.1**	**-0.1**	**4.79**
-0.05	-0.025	-0.0125	0.0	0.0	6.79	0.0	0.0	0.0	-0.05	-0.05	10.01
0.0	0.0	0.0	0.0	0.0	7.99	0.0	0.0	0.0	0.0	0.0	28.48

The camera trajectories in ORB-SLAM [8] with different distortion parameters are given in Fig. 4. The estimate for distortion parameters with error will cause the ORB-SLAM algorithm unstable at the corner. At the same time, we give some specific values of position errors in Table 3. Obviously, we can find that the more accurate distortion parameters are estimated, the smaller the positioning error is achieved. Besides, the error estimate for distortion parameters may directly lead to the fail of the ORB-SLAM algorithm. In general, this experiment demonstrates the virtual camera with the proposed geometric consistency model could be used as a data generator for the development of VSLAM algorithms. The proposed model enables quick testing and evaluation of VSLAM algorithms in a texture-rich environment. In the future, this model can be further developed to test the algorithm under different weather conditions.

5   Conclusion

In this paper, we propose a geometric consistency model of virtual camera for vision-based SLAM simulation and we also provide an efficient implement of our geometric consistency model which can help researchers build custom VSLAM data sets. In the future, we plan to support more camera models like the fisheye camera model.

Acknowledgment. This work was supported by the National Nature Science Foundation of China (NSFC) under Grant 61873163, Grant 61402283, and the Shanghai Science and Technology Committee under Grant 20511103103, and Equipment Pre-Research Field Foundation under Grant 61405180205, Grant 61405180104.

References

1. Bettens, A.M., et al.: UnrealNavigation: simulation software for testing SLAM in virtual reality. In: AIAA Scitech 2020 Forum, p. 1343 (2020)

2. Delmerico, J., Cieslewski, T., Rebecq, H., Faessler, M., Scaramuzza, D.: Are we ready for autonomous drone racing? The UZHFPV drone racing dataset. In: 2019 International Conference on Robotics and Automation (ICRA) (2019)
3. Dosovitskiy, A., Ros, G., Codevilla, F., Lopez, A., Koltun, V.: CARLA: an open urban driving simulator. In: Proceedings of the 1st Annual Conference on Robot Learning, vol. 78, pp. 1–16. PMLR (2017)
4. Drap, P., Lefèvre, J.: An exact formula for calculating inverse radial lens distortions. Sensors **16**(6), 807 (2016)
5. Heikkila, J.: Geometric camera calibration using circular control points. IEEE Trans. Pattern Anal. Mach. Intell. **22**, 1066–1077 (2000)
6. Koenig, N., Howard, A.: Design and use paradigms for Gazebo, an open-source multi-robot simulator. In: Proceedings of 2004 IEEE/RSJ International Conference on Intelligent Robots and Systems, 2004 (IROS 2004) (2004)
7. Merriaux, P., Dupuis, Y., Boutteau, R., Vasseur, P., Savatier, X.: A study of vicon system positioning performance. Sensors **17**(7), 1591 (2017)
8. Mur-Artal, R., Montiel, J.M.M., Tardos, J.D.: ORB-SLAM: a versatile and accurate monocular SLAM system. IEEE Trans. Robot. **31**(5), 1147–1163 (2015)
9. Pei, L., et al.: IVPR: an instant visual place recognition approach based on structural lines in manhattan world. IEEE Trans. Instrum. Meas. **69**(7), 4173–4187 (2019)
10. Qiu, W., et al.: UnrealCV: virtual worlds for computer vision. In: Proceedings of the 25th ACM International Conference on Multimedia (2017)
11. Ricolfe-Viala, C., Sanchez-Salmeron, A.J.: Lens distortion models evaluation. Appl. Opt. **49**(30), 5914–5928 (2010)
12. Shah, S., Dey, D., Lovett, C., Kapoor, A.: AirSim: high-fidelity visual and physical simulation for autonomous vehicles. In: Hutter, M., Siegwart, R. (eds.) Field and Service Robotics. SPAR, vol. 5, pp. 621–635. Springer, Cham (2018). https://doi.org/10.1007/978-3-319-67361-5_40
13. Sturm, J., Engelhard, N., Endres, F., Burgard, W., Cremers, D.: A benchmark for the evaluation of RGB-D SLAM systems. In: 2012 IEEE/RSJ International Conference on Intelligent Robots and Systems (2012)
14. Wen, F., Pei, L., Yang, Y., Yu, W., Liu, P.: Efficient and robust recovery of sparse signal and image using generalized nonconvex regularization. IEEE Trans. Comput. Imaging **3**(4), 566–579 (2017)
15. Weng, J., Cohen, P., Herniou, M.: Camera calibration with distortion models and accuracy evaluation. IEEE Trans. Pattern Anal. Mach. Intell. **10**, 965–980 (1992)
16. Wu, Y., Wu, Y., Gkioxari, G., Tian, Y.: Building generalizable agents with a realistic and rich 3D environment. arXiv preprint arXiv:1801.02209 (2018)
17. Zhu, Z., et al.: Real-time indoor scene reconstruction with RGBD and inertial input. In: Proceeding of 2019 IEEE International Conference on Multimedia and Expo (ICME 2019), Shanghai, China (2019)
18. Zou, D., Wu, Y., Pei, L., Ling, H., Yu, W.: StructVIO: visual-inertial odometry with structural regularity of man-made environments. IEEE Trans. Robot. **35**(4), 999–1013 (2019)

CPI: LiDAR-Camera Extrinsic Calibration Based on Feature Points with Reflection Intensity

Lihao Tao, Ling Pei$^{(\boxtimes)}$, Tao Li, Danping Zou, Qi Wu, and Songpengcheng Xia

Shanghai Key Laboratory of Navigation and Location-Based Services,
School of Electrical Information and Electrical Engineering,
Shanghai Jiao Tong University, Shanghai, China
ling.pei@sjtu.edu.cn

Abstract. Autonomous navigation and unmanned systems are gaining increasingly attention by the research community, especially towards solving problems related to multi-sensory data fusion for navigation and positioning in complex scenarios. In this paper, a novel calibration method and experimental setup are proposed to calibrate the extrinsic parameters between a LiDAR and a camera. This method is based on the directly extracted LiDAR feature points considering reflection intensity and visual feature points derived from ArUco over a calibration board. Therefore multiple calculation points correspondences between the camera and the LiDAR frames are retrieved to achieve an accurate extrinsic calibration. Aiming at proving the quality of the proposed method, setup, process and experimental results are described and commented in this paper. In our experiments we achieved a LiDAR calculation point extraction error of less than 0.25 cm, obtaining an indirect binocular camera extrinsic calibration error over the translation about 0.4 cm and 0.7° error over the rotation. The proposed method compared to the other state-of-the-art method, results improved to more than 3 times on the accuracy and about 5 times on the convergence level. In addition, after transforming the LiDAR frame into the camera frame with our calibration results, the image and the LiDAR point cloud can match and coincide perfectly.

Keywords: Extrinsic calibration · Reflection intensity · LiDAR · Camera

1 Introduction

Data obtained by various types of sensors can meet the needs of navigation and positioning [3] in complex environment where unmanned system is widely used. Multiple sensors such as LiDAR (Light Detection and Ranging) [2], camera, GNSS (Global Navigation Satellite System), IMU (Inertial Measurement Unit) can provide redundant environmental information which can be fused by appropriate algorithm to decrease errors on overall calculation of SLAM (Simultaneous Localization and Mapping) [13,15]. Robust navigation and positioning

© Springer Nature Switzerland AG 2021
X. Meng et al. (Eds.): SpatialDI 2020, LNCS 12567, pp. 281–290, 2021.
https://doi.org/10.1007/978-3-030-69873-7_21

results can be achieved in dynamic environments with multi-sensor system [8]. LiDAR and camera are widely deployed for environment sensing. LiDAR can obtain structural features in wide range scenarios, while camera can extract various texture features. Accurate extrinsic parameters between the two sensors gain particular importance when tightly data fusing of the two sensors [11].

LiDAR points scattered on corners of the calibration board cannot be directly extracted due to the sparsity of the LiDAR point cloud. Previously, several methods are based on extracting LiDAR points scattered on the edges of the calibration board, which can consequently be used to calculate lines by line fitting. Calibration board corner points in LiDAR frame are obtained by calculating the intersections of such lines [14]. These methods contain algorithms of edge extraction [10], line fitting [4] and calculation of intersection, which can bring large errors, with the effect of inaccurate calculation of corner points.

This paper proposes a LiDAR-Camera extrinsic calibration method with feature points direct extraction based on reflection intensity namely CPI (Calibration based on feature Points with reflection Intensity). The main contributions of this paper are:

- A novel LiDAR-Camera calibration board considering LiDAR reflection intensity is proposed. With proposed calibration board, the accurate and direct extraction of LiDAR feature points on the calibration board based on reflection intensity is realized, without the steps of LiDAR edge points extraction and line fitting. The method in this paper can be processed with only one calibration board, instead of multiple boards in the state-of-the-art solution.
- A LiDAR-Camera extrinsic parameters estimation algorithm with high convergence and accuracy is proposed. In this paper, multiple extracted LiDAR feature points are smoothed, multiple final calibration results are averaged, convergent and accurate calibration results are obtained. The convergence and the accuracy of the proposed algorithm improved to about 5 times and more than 3 times respectively, compared to the state-of-the-art algorithm.

2   Related Work

A checkerboard was firstly introduced by Zhang [12] to calibrate the extrinsic parameters between a 2D LiDAR and a camera. Zhang placed a checkerboard with several poses and found that, the LiDAR points scattered on the checkerboard and board plane parameters estimated by camera can construct plane-line correspondences. They optimized for the extrinsic parameters with such correspondences. Manohar Kaul et al. [9] proposed a filtering and lifting framework that augments a standard 2D spatial network model with massive aerial laser scan data to an accurate 3D model. The lifting part of this framework contains triangulation algorithm, which can approximate the surface of the LiDAR point clouds. More recently, a different approach was adopted by Ankit Dhall et al. [6] to calibrate a 3D LiDAR and a camera. This has been widely applied in industrial domain. Dhall proposed an experimental setup which is composed by multiple calibration boards. The method uses ArUco Marker [5] to calculate the calibration board corner points in the camera frame, while LiDAR points scattered on the edges of the calibration board can be extracted to calculate lines by line fitting, the same corner points in the LiDAR frame can be obtained by calculating the intersections of such lines. Calibration board corner points in the camera frame and the LiDAR frame can be used to calculate LiDAR-Camera extrinsic parameters. In practice, the method includes instable edge extraction and line fitting algorithms which bring large errors to the calibration. In addition, the method is complicated because multiple calibration boards are needed. Generally, extrinsic calibration of a camera and a LiDAR needs calibration board to provide sufficient visual and LiDAR features. Therefore, this paper concentrates on a novel calibration approach to extract corresponding LiDAR features and visual features on one calibration board.

3   Methodology Description and Novel Calibration Board

An overview of the notation used in the paper is provided in Table 1. Throughout this paper we use formula (1) to calculate the extrinsic parameters using ICP (Iterative Closest Point) [1,7] algorithm.

Table 1. Notation description

Notation	Description	Notation	Description
E_j	Feature module	P_j^L	LiDAR calculation point
E_j^B	Reference bar	P_j^C	Visual calculation point
E_j^M	LiDAR marker	T_L	LiDAR frame
P_j	Center point	T_C	Camera frame

$$T_L^C = P_j^L \xleftrightarrow{\text{ICP}} P_j^C \tag{1}$$

where transformation T_L^C stands for rotating (R_L^C) and translating (t_L^C) from the LiDAR frame to the camera frame. The novel calibration board proposed in this paper is composed by 26 feature modules and one ArUco Marker, as shown in left side of Fig. 1. Each feature module is composed by one reference bar and one LiDAR marker, as shown in right side of Fig. 1. Center point is the center of the LiDAR marker. The proposed algorithm accurately extracts the LiDAR feature points scattered in the LiDAR marker, then average the LiDAR feature points to obtain LiDAR calculation point, which is equivalent to center point in LiDAR frame. Using ArUco Marker can obtain visual calculation point, which is equivalent to center point in camera frame, more information is detailed in visual calculation points extraction section. LiDAR calculation point and visual calculation point are corresponding to center point on the calibration board, which can provide sufficient constraints for calibration. An accurate and convergent LiDAR-Camera extrinsic parameters estimation algorithm is proposed for these constraints.

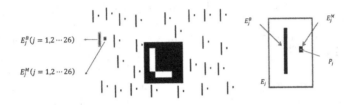

Fig. 1. Proposed calibration board

4   LiDAR Calculation Points Extraction

4.1   LiDAR Feature Points Extraction Based on LiDAR Reflection Intensity

Reflection intensity of LiDAR points is related to the reflectivity of the material. Table 2 shows the statistics of reflection intensity of LiDAR points scattered on four common materials, black PVC tape and calibration board surface have significantly different reflectivity, which indicates that black PVC tape can be the material of the feature module used for LiDAR feature points extraction on the calibration board surface.

Table 2. Reflectivity of common materials

Material	Black PVC tape	Calibration board surface	White paper	Black paper
I	0–50	>75	>90	0–95

As shown in left side of Fig. 2, the pattern with black PVC tape on the calibration board surface can be easily identified in LiDAR point cloud using reflection intensity as shown in right side of Fig. 2, which illustrates the stability of the LiDAR feature points extraction with the two materials.

Fig. 2. Stable LiDAR feature extraction with reflection intensity

Threshold applied on reflection intensity I is set as 50 based on Table 2. An example of such LiDAR feature points extraction is shown in Fig. 3, extracted 3D LiDAR feature points are projected on 2D plane. Features with intensity below 50 this paper are indicated as white pixels and others are indicated as black pixels, as shown in right hand image of Fig. 3.

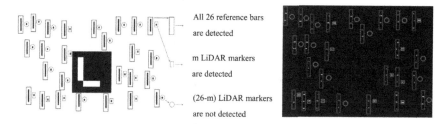

Fig. 3. LiDAR feature points extraction

With the large-sized black PVC tape, the reference bar in a feature module can be easily detected. Using the detected reference bar, corresponding LiDAR marker in the same module can be positioned accurately. Usually, the number of detected LiDAR markers m are less than the number of detected reference bars due to small size of markers missing detected with the sparse LiDAR point cloud. We extract LiDAR feature points scattered in m detected LiDAR markers $E_{j_n}^M (n = 1, 2 \cdots m)$. In the example shown in Fig. 3, all the reference bars and $m = 11$ LiDAR markers have been detected.

4.2   Smooth of Extracted LiDAR Feature Points

A noise LiDAR point can give a wrong coordinate and affect calibration results seriously. In order to improve the robustness of the calibration, our algorithm demand to scan the calibration board N times to increase the number of LiDAR feature points in a LiDAR marker. We find that when $N = 100$, the calibration results are accurate. Using formula (2) to obtain LiDAR calculation point $P_{j_n}^L (n = 1, 2 \cdots m)$, which is equivalent to center point P_{j_n} in LiDAR frame.

$$P_{j_n}^L = \frac{1}{b} \{E_{j_n}^M\} \tag{2}$$

$\{E_{j_n}^M\}$ contains the LiDAR feature points N times scattered in LiDAR marker $E_{j_n}^M$. b is the number of LiDAR feature points in $\{E_{j_n}^M\}$.

5   Visual Calculation Points Extraction

As shown in Fig. 4, computer can calculate transformation T_A^C between camera frame T_C and ArUco Marker frame T_A. We measured center point P_{j_n} on the

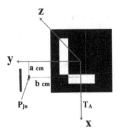

Fig. 4. Measurement of center point on the calibration board

calibration board and obtain $P_{j_n}^A(a, b, 0)$, which is equivalent to center point in ArUco Marker frame T_A.

$$P_{j_n}^C = T_A^C P_{j_n}^A \tag{3}$$

Formula (3) is used to obtain visual calculation point $P_{j_n}^C$, which is equivalent to P_{j_n} in camera frame. For more information regarding ArUco Marker we suggest further reading [5].

6   Extrinsic Parameters Calculation

LiDAR calculation point $P_{j_n}^L$, which is equivalent to center point P_{j_n} in LiDAR frame. Visual calculation point $P_{j_n}^C$, which is equivalent to P_{j_n} in camera frame. Formula (4) describes sufficient constraints for calibration.

$$P_{j_n}^C = T_L^C P_{j_n}^L \, (n = 1, 2 \cdots m) \tag{4}$$

ICP algorithm is used to estimate LiDAR-Camera extrinsic parameters $[R_L^C | t_L^C]$ with such constraints. We calculate Z calibration results and use formula (5) for getting final calibration results with high convergence.

$$\overline{t_L^C} = \frac{1}{Z} \sum_{i=1}^{Z} t_i \qquad r = \frac{1}{Z} \sum_{i=1}^{Z} Q_i \tag{5}$$

In formula (5), t_i is the i_{th} translation result and $\overline{t_L^C}$ is averaged result. We average rotation variable by averaging quaternions. Q_i is the quaternion form of the i_{th} rotation result, r is averaged result in quaternion. We find that when $Z = 20$, the convergence of the calibration results can meet demand.

7   Experimental Verification

7.1   Extraction Accuracy of LiDAR Calculation Point

The extraction accuracy of LiDAR calculation point $P_{j_n}^L$ can directly affect the accuracy of the calibration. In this paper, $D_{j_n}^{j_{n+1}{}'}$ is the distance between two adjacent LiDAR calculation points $P_{j_n}^L$ and $P_{j_{n+1}}^L$. Ground truth $D_{j_n}^{j_{n+1}}$ is the manually measured distance between adjacent two center points P_{j_n} and $P_{j_{n+1}}$ on the calibration board.

$$e_{j_n}^{j_{n+1}} = \sqrt{\|D_{j_n}^{j_{n+1}} - D_{j_n}^{j_{n+1}\prime}\|^2} \qquad \bar{e} = \frac{1}{2m}[(\sum_{n=1}^{m-1} e_{j_n}^{j_{n+1}}) + e_{j_1}^{j_m}] \qquad (6)$$

We use left side of formula (6) to compare $D_{j_n}^{j_{n+1}\prime}$ with ground truth $D_{j_n}^{j_{n+1}}$ and obtain sum extraction error $e_{j_n}^{j_{n+1}}$, which contains the extraction errors of $P_{j_n}^L$ and $P_{j_{n+1}}^L$. The right side of formula (6) is used to calculate the average extraction error $\bar{e}$, which is related to the *size* of the LiDAR marker. We design an experiment made on *size* and the average extraction error of LiDAR calculation point is shown in Table 3.

Table 3. Average extraction error of LiDAR calculation point

size	0.6 cm × 0.6 cm	1 cm × 1 cm	1.4 cm× 1.4 cm	1.8 cm × 1.8 cm
$\bar{e}$ (cm)	**0.24**	**0.28**	0.37	0.46

Table 3 shows that when *size* is smaller than 1 cm × 1 cm, average error is less than 0.25 cm, extraction of LiDAR calculation point has a high accuracy. When *size* is larger than 1 cm × 1 cm, average error increases. The main reason is that, with the increase of *size*, the number of noise LiDAR points in LiDAR marker increases, which can lead to the decrease of the extraction accuracy.

7.2   Quantitative Analysis of Accuracy and Convergence of Calibration Results

Since the positions of LiDAR frame and camera frame are not given by factory, the ground truth of LiDAR-Camera extrinsic parameters cannot be obtained by manual measurement. The extrinsic parameters $T_{C_r}^{C_l\prime}$ between two lenses of a MYNTAI binocular camera given by factory are used as ground truth to evaluate the accuracy of our algorithm. We use right lens calibration results $T_L^{C_r}$ and left lens calibration results $T_L^{C_l}$ to calculate the extrinsic parameters between two lenses of the binocular camera [14]. Formula (7) is used to calculate the calibration results $T_{C_r}^{C_l}$ and error transformation matrix T_e.

$$T_{C_r}^{C_l} = T_L^{C_l}(T_L^{C_r})^{-1} \qquad T_e = T_{C_r}^{C_l}(T_{C_r}^{C_l\prime})^{-1} \qquad (7)$$

T_e contains the calibration errors of $T_L^{C_r}$ and $T_L^{C_l}$. Our algorithm is accurate when T_e is close to zero. We evaluate convergence of the results using RMSE (Root Mean Square Error). *Size* of LiDAR marker, number m of LiDAR calculation points used for calibration, can affect the accuracy and convergence of the calibration results. We design experiments on these factors. Through experiments, when $size = 0.6$ cm $\times$ 0.6 cm, $m = 10$, accuracy and convergence of the calibration results perform best, we set condition above as proposed condition. We compare our method with Dhall's method by real experiments. Experimental setup proposed by Dhall composed by two calibration boards, suspended by a rope in front of the LiDAR and camera about 2 m. Multiple Dhall's calibration results are processed by formula (7) and average errors and RMSE are obtained, which are compared with that from our method under proposed condition, as shown in Table 4.

Table 4. Average errors and RMSE of binocular camera extrinsic calibration results

T_e	x (cm)	y (cm)	z (cm)	sum (cm)	yaw (deg)	$pitch$ (deg)	$roll$ (deg)	sum (deg)
Our's	0.40	0.04	0.09	0.41	0.04	0.48	0.16	0.68
Dhall's	1.32	2.67	0.46	3.01	0.38	−0.02	−1.53	1.93
Our RMSE	0.46	0.40	0.13	0.62	0.04	0.40	0.16	0.60
Dhall's RMSE	1.99	2.74	0.89	3.50	1.21	0.86	1.03	3.10

Table 4 shows that when the calibration with proposed condition, the translation error of Dhall's algorithm is 3.01 cm, which is about 7 times that of our algorithm. The rotation error of Dhall's algorithm is 1.93°, which is about 2.8 times that of our algorithm. The translation and rotation RMSE of Dhall's algorithm are about 5 times that of our algorithm, which illustrates that the convergence of our algorithm is about 5 times that of Dhall's algorithm. When $size = 0.6$ cm $\times$ 0.6 cm, with the LiDAR calculation points number m changing, the errors of calibration results are shown in Table 5.

Table 5. Calibration errors with the calculation points number changing

m/T_e	x (cm)	y (cm)	z (cm)	sum (cm)	yaw (deg)	$pitch$ (deg)	$roll$ (deg)	sum (deg)
5	−1.82	1.14	0.38	2.18	0.40	−0.35	−0.89	1.64
6	0.86	−0.05	0.09	0.87	−0.26	−0.43	0.37	1.06
7	0.09	0.53	0.29	0.61	0.10	−0.27	−0.77	1.14
8	−0.68	0.02	0.09	0.69	0.72	0.07	0.13	0.92
9	0.32	0.36	0.15	0.50	0.34	0.14	−0.45	0.93
10	0.40	0.04	0.09	0.41	0.04	0.48	0.16	0.68

Table 5 shows that the calibration errors are extremely large when m is less than 6, calibration errors decrease with the increase of m. The main reason is that m is the number of constraints in the calibration, the calibration results are accurate when constraints are sufficient. When $m = 9$, with the LiDAR marker $size$ changing, the calibration errors are shown in Table 6. Table 6 shows that the accuracy of calibration results is best when $size$ is 0.6 cm $\times$ 0.6 cm, with the translation error less than 0.5 cm, rotation error less than 1°. Translation and rotation errors increase with the *size* increasing. The main reason is that, *size* is related to LiDAR calculation points extraction.

Table 6. Calibration errors with the LiDAR marker *size* changing

$size/T_e$	x (cm)	y (cm)	z (cm)	sum (cm)	yaw (deg)	$pitch$ (deg)	$roll$ (deg)	sum (deg)
0.6 cm × 0.6 cm	**0.32**	**0.35**	**0.15**	**0.50**	**0.33**	**0.13**	**−0.45**	**0.91**
1 cm × 1 cm	0.75	0.10	0.04	0.76	−0.36	0.76	0.58	1.70
1.4 cm × 1.4 cm	−0.42	−0.54	1.15	1.34	−0.69	0.52	0.24	1.45
1.8 cm × 1.8 cm	−1.38	1.12	−0.06	1.78	0.30	−0.98	−0.85	2.13

7.3   Visual Evaluation of the Calibration Results

After transforming the LiDAR frame into the camera frame with our calibration results, the image and the LiDAR point cloud can match and coincide perfectly, which proves the correctness of our method (Fig. 5).

Fig. 5. Left side is from our method, right side is from Dhall's method

8   Conclusion

A LiDAR-Camera extrinsic calibration method with points feature extraction based on reflection intensity namely CPI is proposed in this paper. An algorithm which can directly extract LiDAR feature points is introduced, with a novel calibration board, the accurate and direct extraction of LiDAR feature points on the calibration board is realized. All the calibration procedures can be completed with one calibration board is another novelty in this paper. A LiDAR-Camera extrinsic parameters calculation algorithm with high convergence is proposed, the calibration results show better performances in terms of convergence and accuracy than the state-of-the-art method. The experiment results show that the error of LiDAR calculation point extraction is within 0.25 cm. Using our method to calibrate the two lenses of a MYNTAI binocular camera, the error of translation is about 0.4 cm, the error of rotation is within 0.7°, the RMSE of the results is small, which improved to more than 3 times on the accuracy level and about 5 times on the convergence level compared with the state-of-the-art method. Similar experiments are performed with two Point Grey cameras, calibration results show similar accuracy and convergence compared with that from MYNTAI binocular camera, which explains the robustness of our method.

Acknowledgement. This work was supported by the National Nature Science Foundation of China (NSFC) under Grant 61873163, Equipment Pre-Research Field Foundation under Grant 61405180205, the Shanghai Science and Technology Committee under Grant 20511103103.

References

1. Armesto, L., Minguez, J., Montesano, L.: A generalization of the metric-based iterative closest point technique for 3D scan matching. In: 2010 IEEE International Conference on Robotics and Automation, pp. 1367–1372. IEEE (2010)
2. Chen, C., et al.: Trajectory optimization of LiDAR SLAM based on local pose graph. In: Sun, J., Yang, C., Yang, Y. (eds.) CSNC 2019. LNEE, vol. 562, pp. 360–370. Springer, Singapore (2019). https://doi.org/10.1007/978-981-13-7751-8_36
3. Chen, Y., et al.: Knowledge-based error detection and correction method of a multi-sensor multi-network positioning platform for pedestrian indoor navigation. In: IEEE/ION Position, Location and Navigation Symposium, pp. 873–879. IEEE (2010)
4. Choi, S., Kim, T., Yu, W.: Performance evaluation of RANSAC family. J. Comput. Vision **24**(3), 271–300 (1997)
5. A. de la Visión Artificial: ArUco. a minimal library for augmented reality applications based on opencv. Dosegljivo (2015). http://www.uco.es/investiga/grupos/ava/node/26. Dostopano 16 Apr 2016
6. Dhall, A., Chelani, K., Radhakrishnan, V., Krishna, K.M.: LiDAR-camera calibration using 3D–3D point correspondences. arXiv preprint arXiv:1705.09785 (2017)
7. Du, S., Zheng, N., Ying, S., Liu, J.: Affine iterative closest point algorithm for point set registration. Pattern Recogn. Lett. **31**(9), 791–799 (2010)
8. Jiang, W., Xu, C., Pei, L., Yu, W.: Multidimensional scaling-based TDOA localization scheme using an auxiliary line. IEEE Signal Process. Lett. **23**(4), 546–550 (2016)
9. Kaul, M., Yang, B., Jensen, C.S.: Building accurate 3D spatial networks to enable next generation intelligent transportation systems. In: 2013 IEEE 14th International Conference on Mobile Data Management, vol. 1, pp. 137–146. IEEE (2013)
10. Xia, S., Wang, R.: A fast edge extraction method for mobile lidar point clouds. IEEE Geosci. Remote Sens. Lett. **14**(8), 1288–1292 (2017)
11. Zhang, J., Singh, S.: Visual-lidar odometry and mapping: low-drift, robust, and fast. In: 2015 IEEE International Conference on Robotics and Automation (ICRA), pp. 2174–2181. IEEE (2015)
12. Zhang, Q., Pless, R.: Extrinsic calibration of a camera and laser range finder (improves camera calibration). In: 2004 IEEE/RSJ International Conference on Intelligent Robots and Systems (IROS) (IEEE Cat. No. 04CH37566), vol. 3, pp. 2301–2306. IEEE (2004)
13. Zhou, H., Zou, D., Pei, L., Ying, R., Liu, P., Structslam, W.Y.: Visual slam with building structure lines. IEEE Trans. Veh. Technol. **64**(4), 1364–1375 (2015)
14. Zhou, L., Li, Z., Kaess, M.: Automatic extrinsic calibration of a camera and a 3D LiDAR using line and plane correspondences. In 2018 IEEE/RSJ International Conference on Intelligent Robots and Systems (IROS), pp. 5562–5569. IEEE (2018)
15. Zou, D., Wu, Y., Pei, L., Ling, H., Yu, W.: StructVIO: visual-inertial odometry with structural regularity of man-made environments. IEEE Trans. Rob. **35**(4), 999–1013 (2019)

Author Index

Printed in the United States
By Bookmasters